D1020098

A Dark Place in the Jungle

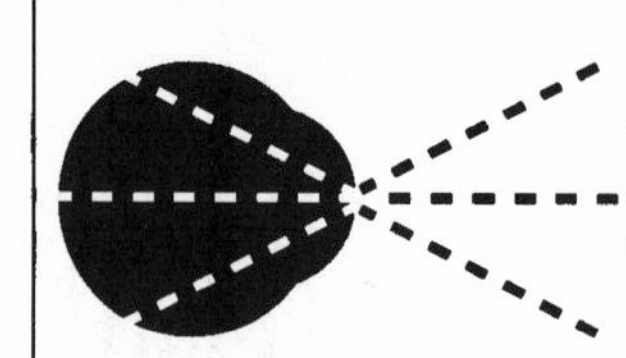

A DARK PLACE IN THE JUNGLE

Linda Spalding

G.K. Hall & Co. • Thorndike, Maine

First published in a somewhat different form in Canada by Key Porter Books Ltd under the title: *The Follow.*
Algonquin edition ©1999 by Linda Spalding.

Published in 2000 by arrangement with Algonquin Books of Chapel Hill, a division of Workman Publishing Co., Inc.

G.K. Hall Large Print Core Series.

The text of this Large Print edition is unabridged.
Other aspects of the book may vary from the original edition.

Set in 16 pt. Plantin by Rick Gundberg.

Printed in the United States on permanent paper.

Library of Congress Cataloging-in-Publication Data

Spalding, Linda.
A dark place in the jungle / by Linda Spalding.
p. cm.
Originally published: Chapel Hill : Algonquin, © 1999.
ISBN 0-7838-8967-4 (lg. print : hc : alk. paper)
1. Orangutan — Behavior — Research — Borneo. 2. Spalding, Linda — Journeys — Borneo. 3. Galdikas, Birutâ Marija Filomena. 4. Borneo — Description and travel. I. Title.
QL737.P96 S62 2000
599.88′315′095985—dc21 99-086319

For *Edith Senner Dickinson*

We cast a shadow on something wherever we stand, and it is no good moving from place to place to save things; because the shadow always follows.

— E. M. FORSTER

CONTENTS

INDONESIA
N
VIETNAM
Ho Chi Minh City
SOUTH CHINA SEA
PHILIPPINES
PALAWAN
MALAYSIA
Singapore
BORNEO
Tanjung Puting
National Park
CELEBES
(Sulawesi)
SUMATRA
JAVA SEA
Semarang
JAVA
BALI
Bogor
Yogyakarta
Denpasar

Prelude

I am standing on a dock in Borneo when my point of view changes. From this moment, everything seems to be covered in a new color, one I have not yet learned to name.

First, I hear Kristin calling. Quickly. She's my daughter, and the urgency in her voice is compelling, although I can hear another, smaller scream as well. Then I'm on the dock looking down at the water and weeds of the swamp, where there is a red-patterned snake with a frog in her mouth. It's the frog who is crying; the sound is small but heartrending. No wonder. The snake's jaws are locked around the frog's flat foot. They're chewing. The foot is disappearing. At this rate, it will take a long time, but as the foot goes, so will go the leg, then the rest of the frog.

There are two men on the dock. One of them is Yadi, the driver of our boat, who refuses to get involved with the snake. Poisonous, he tells us. Very poisonous. The other man works at the lodge at the end of this dock and is enjoying a break from his duties. We've all been relaxing in the way people do who are more or less ignoring each other, but the frog cries frantically. The snake continues to swallow. I'm paralyzed. I remember watching my other daughter, Esta, almost drown and being paralyzed then, too, looking down into the water in the same way: frozen. Who am I to deny the snake her lunch? And how interfere with the fate of a frog who is suddenly quiet, as if resigned to being eaten alive?

Who am I to interfere even in the way the local people on the dock shrug off a cry of pain?

The young man is amused by our interest, and when he leaps up with a piece of lumber in his hand, I suddenly fear for the snake, who has earned her lunch the hard way and who is uncoiled in the underbrush, swallowing trustingly as we stand above her on the dock. Then the chunk of lumber comes down hard, cutting through brush and just missing the snake, who coughs up the frog and slides away. We can see the frog, with a glob of blood on its back, shove itself into the brown river.

This is the moment when the unnameable color saturates water and trees and the rice padis lying beyond them where the rainforest has been cut away in order that people may eat. Even the sky, with its eagles and kingfishers, is a stranger hue.

Until now, I have understood myself.

And most things.

Borneo? It was only a suggestion by someone I'd never met, made while describing the unusual life of a woman who committed herself to a malaria-infested swamp and a group of endangered orangutans. "One of Leakey's three angels," she said. "And a great primatologist. If you're interested in the implications of motherhood, think about adopting an orangutan.

That's what she does."

I was finishing my second novel and the caller was a publisher. I was flattered, even tempted. The thought of the woman and the swamp and the forest around them reawakened thoughts that had been tormenting me for years. Thoughts about our connection to nature. We seem to be wandering outside it, but how can that be? Aren't we made of the same coils of DNA as everything living? Aren't our closest relatives the other great apes (chimpanzees, bonobos, gorillas, and orangutans)? Now only orangutans still live in the trees from whence we came, wandering like nomads through the canopy, without permanent nests, the way we must have wandered once upon a time. Was it settlement that cut us off from nature? Are we human because we left paradise?

Orangutans were virtually ignored by naturalists until a young Canadian named Biruté Galdikas climbed out of a sampan in 1971 and settled in an abandoned forester's hut in Borneo's nature reserve of Tanjung Puting. As an anthropologist, she had a mission to study the tree dweller and to learn through this study more about the origins of human behavior. In those days, Biruté — like Louis Leakey's other protégées, Jane Goodall and Dian Fossey — "followed" a primate for days, even weeks, learning what she ate, where she slept, how she raised her children, what mark her life in the forest made. It was a new kind of research, observation in situ, one at which Leakey found women particularly adept. Perhaps he had not counted on the mothering for which Biruté became famous, but he assumed that great apes, who are hierarchical, would be less threatened by a woman than by a man.

I knew nothing about orangutans or the rainforest where they live, but I knew enough to know it was going to be hard to follow Biruté. The trail was overgrown. And I am no athlete. Still, Rousseau's prescription for us, who have drifted so far from our origins, was to make two journeys: one to a place where life is still uncorrupted, and another into the self. And Biruté had written in her memoir, *Reflections of Eden*, "Every trip into the field is also a journey into yourself."

By following Biruté, perhaps I could accomplish both journeys.

My follow began in the north, in Toronto, where I live, but it took me deep into the rainforest and even deeper, into a tangled web of relationships and beliefs. I met snakes, frogs, scientists, and the man convicted of smuggling a group of baby orangutans called the Bangkok Six. I met the Dayak (the historical headhunters of Borneo), government officials, passionate ecotourists, and plenty of troubled orangutans. Each of them was a station along my way. Every footstep affects the forest, but a breath taken in or given out changes nothing, until the breath is that of *Homo sapiens*. Until the breath is human, all that changes is the weather, the level of the damp, the season. There is swelling and subsiding, birth and death. Only we, the human finger on the primate hand, desire to hold, to save, to change.

Biruté and I are both children of the magical sixties, formed by that time when everything seemed possible, and the end justified the means. We no longer speak the same language. We chose different ways. But we have much in common. I knew little about the new theories of our primate heritage, but those thoughts about nature and our place in it had been nagging at me. I was fifty and I was willing to see that as a defining moment. Even if Biruté is as elusive as the orangutans she follows, I thought, she might lead me to an understanding of how we *Homo sapiens* got ourselves thrown out of the garden and how we must look, in our exile, to the many eyes watching from the trees. Of course there was something else. A quest is as much about the seeker as the sought. We look in the mirror of another face to find ourselves. This is a signature act of all primates.

Picking and Saving

They came to my hide and their piercing brown eyes met my grey ones: scratching themselves with wonder, they then walked away without having solved the mystery.

— Adriaan Kortlandt

"One of the things I've learned," Biruté tells a college audience in Long Beach, California, in 1995, "is that orangutans have the most intense mother-offspring relationship of probably any mammal. If you look at the somewhat solitary life of the orangutan in adulthood, what prepares him or her for that life is the very intense period, which lasts seven to eight — usually about ten — years, when the immature orangutan, male or female, is in contact with his or her mother one hundred percent of the time."

I am sitting in a room with many others, thinking about my own somewhat solitary past. Once, I also had an intense mother-offspring relationship, forming a unit with my two daughters and moving from one nest to another for several years. Twelve, to be exact. Twelve years. This is not to imply that I was chaste, but the men I knew were never fathers to my children, never partners in my life, and there was bewilderment for all of us in that.

For twelve years, I was a single mother living in a series of small houses on Oahu. One of my daughters was born there. Both lived within earshot of the ocean, under the broken sides of volcanoes, beneath the spread of flowering and fruiting trees. It was a dreamlike, intensely intimate time, and even as the tides pulled and pushed at the shores around us, the earth seemed careless; it was inappropriately fecund. Volcanoes erupted, the lava cooled. Out of this lava, trees sprang up and bloomed. Around us there was abundance, but the scars of certain hardships, insecurities, and traumas are permanent. As my daughters have become adults, the mysteries of their joined childhoods have become more compelling to them, and I have felt, at times, like an object of study — someone whose life is already being sifted and analyzed as if it is finished and done. Things that seem to me to be still in process are, for them, parts of a remote and reducible past. It's as if, leaving Hawaii, we left solid traces behind, fossilized but decipherable.

"Conservationists, ecologists must become political," Biruté is saying. "Global forces . . ." I must have missed something. My habit of drifting off into self-reflection will have to stop if I am going to attend to Biruté's work and life. After all, I'm sitting in the dark room because I have begun a *follow,* a journey no more surreal than many others I have made, but one I feel completely unprepared for. I'll be entering this stranger's territory. Present and past.

"Do you want to come to Borneo?" I would ask my girls the following week, because the more I thought about making the journey, the more important it seemed to take them with me. I would tell them I was on the trail of something important. A follow, as I would explain it, is a form of research in which the subject is observed from a distance, and in which a record is kept of the surroundings, noting moods of subject as well as environment. But this would be something else as well. It would be a reunion of our original trinity, since, in the upheaval of moving to Canada for love, of blending two families, I had not been alone with them for more than a decade. They were living away from home, and I was still trying to adjust.

I'd buy the tickets and pay for room and board but nothing else. No goodies, no extras, because part of our family history is my maternal fear of inequality. (Is this a "natural" maternal fear or was it a result of single parenting? I was morbidly worried lest I cast more light on one child than the other.) We would introduce ourselves to Indonesia gradually, beginning in Bali, which is perhaps most felicitous, and moving up and through the archipelago in order to get a feel for the culture and in order to spend the two weeks Biruté insists must be spent in Asia before meeting the endangered orangutans. Bali and Java would be our quarantine.

Already I was hatching my plan.

Long before I'd flown to L.A., I'd written to the Orangutan Foundation International and become a member. There are several OFI branches around the world, but the main office is in Los Angeles, and since it's an organization started by Biruté and basically set up to support her, L.A. seemed like the right starting place. I'd been doing my homework and reading about Louis Leakey and his lifelong search for early man. It was this search that led him to send his three "angels" — Goodall, Fossey, and Galdikas — off into the wilds of Africa and Borneo to study the great apes, reasoning that their physical and social environments might shed light on the lives and adaptations of the first hominids. I'd also been talking to people who had worked at Camp Leakey, Biruté's research station in Borneo, for periods ranging from two weeks to two years. They told me they had stayed in a "team house." They had gone as students, as scientists, or as volunteers with Earthwatch, an organization dedicated to helping ecological trouble spots around the world. The volunteers had spent about $2000 for the privilege of being part of the Camp Leakey scene, in addition to the cost of airfare, which is also about $2000. Assignments usually involved searching for wild orangutans, then following them while collecting data on movement, activity, food, and so forth. I met a Toronto woman named Connie

Russell, who had gone as a student, working on a study of ecotourism. "This is not the fun-in-the-sun type of travel," she had written in her report, "nor does this reflect most ecotourism adventures. Instead, these people offer to assist with the project's activities in return for food, accommodation, field training, lectures by Galdikas and other visiting scientists and, of course, the opportunity to 'do science' and 'do good' at the same time."

Doing good seemed to be the common motive.

There were stories of discomfort, stories of not enough food. But in spite of these warnings, I was more and more determined to see the orangutans, the rainforest, and Camp Leakey, the last great site of primate research established by a woman and still in her charge. Discomfort didn't worry me. After all, there were people around me who had been there and survived, and all of them remembered the orangutans with pangs of intense nostalgia. "There's no place in the world like Camp Leakey," I'd been told, "where you can live with one of the great apes as an equal." (Among the OFI materials I'd been sent was a set of rules for visitors to the camp. *Rule #9: Remember that in camp the orangutans come FIRST, science second, local staff and people third and we, the foreign researchers, LAST.*) I'd also been told that in 1994 Earthwatch had, for some reason, dropped Biruté and Camp Leakey from its roster of tours and projects, but there was another alternative. After trying to run tours themselves for a summer, OFI had found a travel company in Colorado called Bolder Adventures to manage the organization's tours. So I called them and spoke to someone named Eric. I said I wanted to take my daughters to Borneo to see orangutans.

Eric said that there were two versions of the OFI trip available, with varying amounts of time in Kalimantan — Indonesian Borneo. The cost of the one I wanted was exactly what the Earthwatch trip had cost, a hefty $2000 a person. With airfare another $2000 each, it began to look as if my get-acquainted visit to Borneo was going to cost $12,000 for the three of us, not counting any equipment we might need.

Then I called the OFI. "What about doing this on my own?" I asked.

"You can't get into the park without a guide, and that's what we provide," said the OFI vice-president and longtime Galdikas colleague, Gary Shapiro, who was in the office that day. "If you want to get to Camp Leakey, you have to go with us."

That was disheartening. But it was around this time that I met Dr. Anne Russon, a York University psychologist doing a long-term study of learning and imitation among Borneo orangutans. She said she didn't know of any other tours. "Unless you want to just get a ticket and go," she said. "It's a national park. All you need is a permit."

Which would save $6,000.

I wrote to Biruté, then spoke to her by phone in Vancouver, where she teaches at Simon Fraser University for part of every year. She sounded rushed. "I'm on my way out the door. I'm on my way to Indonesia . . . ," which made me feel very lucky to have caught her at all. "You can write to me in Pasir Panjang," she said. "Or call my mother in L.A. She knows my schedule."

When I spoke to Mrs. Galdikas, *mère,* who lives in Los Angeles, but who once lived in Toronto as I do, she said my best chance of meeting Biruté was in Los Angeles during a promotional tour for her book *Reflections of Eden.* Filomena Galdikas came to Canada as an immigrant with her husband and first child, Biruté, born in 1946. There are tales of Biruté's Lithuanian upbringing in the vicinity of the murky Grenadier Pond in High Park, a site of inspiration for the future scientist, but Toronto is full of such immigrant tales.

In February 1995, I'd flown to L.A. and checked in to an airport hotel. Orangutan time. Sitting still. Waiting. It was the first morning, the first day of my follow.

Biruté was busy. She was moving. Out of the night nest into the trees. She couldn't come to the phone. Once I heard her laughing in the background when I called, but the person speaking — her publicist — said she was engaged. I was invited by this publicist to attend her public readings. "If you want to see Dr. Galdikas, that's where she'll be." In my hotel room, I watched *the People vs. O.J.* The trial was also in its first few days. In the hotel gym, all the TV sets were tuned to it. Outside, on the sidewalk, O.J. T-shirts were for sale and placards sang his praises.

Biruté says we are social animals. She says she has come to terms with this. After years of working with orangutans, she realizes that, unlike them, we need each other more than anything else. We're social creatures who have lost our place in nature, and we depend on the information we acquire from each other in order to survive in unnatural places. Like Los Angeles, where oil rigs shudder up and down the length of the freeways, lapping up the black gold underneath, and an ominous haze covers the sky. But there are pockets of civility here and there, like Brentwood, where Biruté installed her fund-raising flagship, the OFI, and where Nicole Simpson lived with her two children until she was murdered on her own doorstep. I drove around Brentwood one afternoon with a friend and saw Nicole's house and O.J.'s mansion, and a few blocks away we found the offices of the OFI in the pretty house where Biruté's mother lives. I didn't go in. Biruté's publicist had told me the office wasn't open to the public.

"I'm a member," I'd said.

"If you want to see Dr. Galdikas, she's giving a reading in Manhattan Beach."

Which is how I came to find myself in the first

of several crowded rooms watching Birutė and already pondering my own solitary past, although what researchers describe as "solitary" is often a single mother's life. Children don't seem to count. But they are company. They need and they give. These were my thoughts as I sat watching Birutė, who was softer than I had expected, wearing pants and a big shirt. Perhaps because of her large glasses, I couldn't read her eyes.

When she sat down at a table to sign copies of her book, I waited in line and then introduced myself, holding out my copy of *Reflections of Eden*. "Oh," she said, when I gave my name, "you're here."

"I called you when I arrived," I reminded her. "I've been trying to get in touch with you for three days."

She smiled briefly, then signed.

"And I wrote to you," I said. "Remember? I called you in Vancouver and we talked about meeting down here. Then I wrote."

"Oh yes," she said. "That's right."

When I told her I had enjoyed the reading and suggested that she should include her old hometown, Toronto, in her promotional trip, she said, "You could help with that," and introduced me to her agent.

Outside, Birutė's daughter, Jane, was jumping rope on the sidewalk. Inside, her son Binti was sitting close to the door. I knew about them from her book — one by the first husband, one by the second. A third child, Freddy, was in Borneo. "He stays there to help his father, a Dayak farmer," Biruté explained to her audience, and I imagined the two males knee-deep in a rice padi and considered the implications of splitting up children by traditional gender roles. In her talk she had focused largely on her adoption of Indonesian ways.

Binti was talking to someone and I tried unsuccessfully to eavesdrop. Was this what my mission required? I bought a glass of juice and introduced myself to Gary Shapiro. His handshake was cool. We may be social animals, I thought, but like orangutans, we have silent ways of communicating.

The publicist, Nancy Briggs, who teaches at UCLA, had promised me lunch with Biruté if I could get to the Long Beach campus for a morning lecture two days later, so, under the drone of arriving and departing planes, I forced myself awake, struggled onto the freeway again, and sped south in my rented car, finding the right turnoff, the right campus, the right building, even the right classroom — easily recognizable because someone in an orangutan suit was lurking at the back. It was the same orangutan suit I had seen at Biruté's second reading the night before, during which she had again talked about her work, read a selection from her book, and signed copies of it for people who stood in line. "I'm having lunch with you tomorrow," I had said then, to announce my reprieve.

"Oh! We have the mascot from the Orangutan Foundation here," Professor Briggs cooed now, as the furry suit felt its way down one of the aisles and waved to the students as if they were children in need of comic relief. "He represents a real male orangutan. From Borneo! You can't find him? Come on, you know he's very shy and very solitary! Come on! You have to help him up on the stage!" She turned dramatically to Biruté, who had joined her on the platform. After introducing her with obvious admiration, she said, "If you wish to visit her in Borneo, you could call her at 1-800-ORANGUTAN and make arrangements to go. Also, in this month's *Cosmopolitan* there's a marvelous article on Dr. Galdikas and her new book. Last month she had an article in *Discovery* magazine, and last week she was on Connie Chung's *Eye to Eye.* She was listed as 'somebody you should know.' "

When the professor told her students the events that were scheduled around Biruté's campus visit — eleven o'clock lecture, twelve o'clock book signing, one o'clock "brand-new, outstanding version of a video that was done for Dr. Galdikas's Orangutan Foundation with magnificent music. I think you'll enjoy it . . ." — I began to worry about our lunch.

This lecture put Biruté on familiar turf, in front of a college classroom. "I teach at a university in British Columbia, a province of Canada," she told the students, "and the situation in British Columbia, which depends on resources —

forests, mines, et cetera — to fuel the economy, is frighteningly similar to what I see in Borneo." The turf was familiar for another reason: Biruté had done her undergraduate and graduate work nearby, at UCLA. It was there that she heard the lecture by Louis Leakey that seems to have determined the course of her life. Now she said she was going to show slides of Camp Leakey, her research site, and later in the afternoon there would be the video with magnificent music. "For the last twenty-four years, I have been studying orangutans," she said brightly, "and I've been trying to aid the conservation of orangutans in their tropical rainforest habitat. You cannot save orangutans without saving their habitat. The two are totally one . . ." Then she said, pushing her glasses up on her nose, "Somebody asked me who is the orangutans' closest living relative. I had to think about it very hard. But the humans are."

I listened closely, looking around at the students, who were also listening. Some of them had their mouths as well as their ears open, as if they were waiting to bite down on something solid. "We humans, with chimpanzees and gorillas — who are the African apes — form a little genetic group. Humans follow the African apes, so we have those two groups. Basically, we are as related to orangutans as chimpanzees are related to orangutans." I wondered what Leakey would have made of Biruté's explanation of our genetic heritage, the study of which had absorbed him for half a century, causing him to begin the first research into the lives of modern apes.

When the lights went out for her slides, Biruté carried on calmly, obviously familiar with her material, most of which involved pictures taken by her ex-husband, Rod Brindamour. They showed a Biruté who was young and lithe and long-haired, a Biruté captivating in the captivation she felt for the forest orangutans. She looked familiar, like someone I might have known a few years before when we were both married to photographers, running away to islands, chasing dreams. We have that in common, I thought, along with hundreds of provocative images created by those ex-husbands, images we no longer fit.

Then I thought, maybe what we have in common goes even deeper than that.

"The most commonly used vocalization in the wild orangutan," Biruté announced, "is the long call, which the adult male gives. The long call is a very complex vocalization. It begins almost like an opera singer's clearing of the throat, and then it rises into a series of bellows. If you hear these from a distance, and you don't know what it is, you could actually be frightened. It sounds like a drunk elephant going crazy. Or a lion. I haven't heard lions in the wild, but I would say that a male orangutan's bellows compare favorably, in terms of intimidation."

I stared up at the screen and tried to will myself into that faraway place.

It would be tropical, full of life and death and

rot. It would be nature undisturbed. The original place.

"Have you read her doctoral thesis?" Nancy Briggs asked me later, as we regained the sweet outdoors and made our way to the parking lot. "Your time might be better spent doing that."

A small group of students were straggling along with us, and I didn't answer. The lunch had come to nothing. Biruté had stood in a hallway, munching a sandwich, busy signing books. During the video presentation, she'd asked me to hold the leftover bits in a plastic box, and I'd continued to hold them while Biruté posed for photographs with the students. Now I was trying to hear her answer to one of their questions: "Who's going to play you in the movie of your life?"

"Sharon Stone, I think," said Biruté.

An OFI volunteer, who was carrying the orangutan costume in a cardboard box, said that was a perfect choice and I smiled. Biruté had asked me to drive her back to L.A. I was going to talk to her alone. By myself.

In the parking lot, she bade the others goodbye.

"Biruté," Nancy Briggs said, with a ring of authority in her voice, "you come with me." To me she said, "I'm taking her to ABC," then added, "and there's not going to be a book, if that's what you're writing. She's already done the story of her life."

"I came all this way," I said lamely. Then I

tried again; because, after all, I *had* come all that way. "What about tonight?"

"I have to help Jane. With her homework," Birute said, and I thought I detected a shred of helplessness. "And Binti . . . I promised I'd take him to Tower Records . . ."

"I could pick you both up at your mother's and take you wherever you want to go."

"Well, okay." I noticed that she avoided the eyes of her friend and publicist, and I thought I saw Nancy Briggs gritting her teeth. But the episode was unsettling. A woman in charge of another woman's public image had exhibited what my mother would call "bad manners." Biruté, the intrepid explorer, had seemed passive, and most unsettling of all, I had taken what wasn't offered. I had been aggressive and pushy. What had come over me?

Nevertheless, by 7:00 P.M., on my way to Biruté's mother's house, the vast system of freeways, oil rigs and all, seemed again lit by possibilities. I found the address and pulled up at the curb exactly as Biruté and Nancy Briggs arrived. It had been a long day, but inside *mère* Galdikas was waiting, blond, wearing a kerchief, very tall and vigorous; a handsome woman bearing no resemblance to her darker daughter. "You shouldn't have come," she told me, as I appeared in the doorway with Biruté. "My daughter is tired. Can't you see?"

"I'm leaving tomorrow," I said, hoping that she would revert to the hospitality she had seemed to extend over the phone when I'd called from Toronto. At that time she had mentioned a nearby motel and had asked me to bring her a map of Toronto — "for my son," she had said. "I brought your map," I announced, but she pounced on a pile of mail, ripping open envelopes and stuffing them into the trash as if she would be happy to do the same to me.

Biruté went off to see Binti and Jane, and I drank the cup of tea offered by an office volunteer. "Binti's dying for tacos," Biruté said when she reappeared. She was obviously tired, but promises had been made. Hers to Binti. Hers to me. I had eaten at my hotel, but what matter?

I said I would love to take them to dinner. I said I loved Mexican food. Then I led Biruté and her children out to the little Geo and drove away with them, astonished at my sudden good fortune. Brentwood glowed. Instead of trailing through Tower Records, we were going to sit down at a table. We would have a drink, conversation, rapport. It was the fourth day of my follow and I was about to have an encounter. What more could I ask? We drove up Wilshire Boulevard, and suddenly Jane shouted, "Mom, Mom! Look! It's a Pic & Save . . ."

Biruté said, "Would you mind stopping? I haven't been in a store since I arrived. We need so many things."

In we trailed, eighteen-year-old Binti wearing his headset, ten-year-old Jane prancing off toward the stuffed animals — most of which were rabbits in early honor of Easter — and Biruté grabbing a large shopping cart. I grabbed one, too. But I was struggling. Not to shop with Biruté at Pic & Save might seem standoffish. To shop, on the other hand, seemed unserious. Had I come out with her to look for a stuffed rabbit or to discuss the interest I had in her work with an endangered species?

"Made in China," I read aloud, wondering if she would think me shallow, buying useless ceramic elephants from the Third World when the whole planet was in crisis. "If I buy it, it's sure to break before I get it home." Easter decorations dripped off every surface. Hundreds of bunnies. Carrot-shaped baskets. Carrot-shaped teapots. Carrot-shaped pillows and spoons. Picking means making a choice. You take the basket, I'll take the elephant. But there are two kinds of saving. We rescue things or we lock them away. Orangutans. Money.

"Listen, I'd really like to talk to you about Borneo," I said. "And your work." I was going to mention some of the ideas I had about methods of mothering, interspecies adoption and so on, but the aisle of Pic & Save seemed like the wrong place to do it. Child abuse. Abortion. Nature versus nurture.

"What about Borneo?"

"Well, I'd like to come out there. As soon as I can."

"You can go this summer with an OFI group."

"You won't be there this summer; you'll be in Canada."

"That doesn't matter. Someone will be there. Someone just as interesting as I."

Jane eyed me as if she was wondering whether I was good for a stuffed rabbit. I thought about it. I was tempted, but it seemed unfair. "I'd like to come out this winter," I said.

Biruté said, "It's against our rules. We only let people come in teams. We've had problems."

We paid our separate bills and trundled up the sidewalk with my elephant and Biruté's cart, which contained bath mats, shower curtains, picture frames, and a chin-up bar three feet long. Ahead, the Geo looked small and unreliable. How had I had the nerve to pack this valuable woman and her children in it and drive them off into the dangerous Brentwood night? How would I ever pack all the goods we'd bought in its interior, and later, the four of us, duly fattened up on tacos? Biruté grabbed my elbow and steered me away from a dark pile on the sidewalk. When I looked down, I saw that it was a man asleep in a bag. We opened the car and stowed our haul. Biruté asked Binti to return the cart. "Wasn't it nice of them to trust us with it?" she said.

Moments later, we slid into a booth — the only patrons of a very ordinary Mexican restaurant a few doors from Pic & Save. What with the aroma of fried corn and the sharp smell of salsa, I felt an immediate craving for lime and salt and tequila, but Biruté ordered a Coke and her children followed suit, so I restrained myself, still worried about the impression I was making. I asked for iced tea. At that, Biruté and Jane immediately changed their drink orders, which seemed, in its small way, to bode well. They ordered two meals each, although Jane had begun to cough and complained of feeling sick. Indeed, as the evening progressed, she seemed truly ill. The cough got worse. Binti had a headache. "Maybe you're just hungry," Biruté said.

Small and slight, almost fragile, Binti had long, straight dishwater hair and a long fingernail on his left hand. He plays the guitar. He lives alone in Vancouver, having left his father and stepmother and several half siblings in Australia and having rejected his mother's invitation to live with her new family in Kalimantan. Having, I suppose, to live down or up to the famous image of himself as a toddler bathing with an infant orangutan on the cover of *National Geographic*, when Biruté was becoming famous. I'd heard that he wasn't interested in orangutans, but there he was with his mother every evening, and what's more, I'd received a packet of materials from the Orangutan Foundation that included a newsletter for young people edited by Binti Brindamour. "Have you been to Toronto?" I asked him.

"No. What's the scene like?"

"It's great. I think you'd like it." Then I added, "Although that's just a guess, since I really don't know much about what you like."

"Cool. I'd like to see it a lot."

Jane groaned audibly. "Mom, my throat hurts. I don't think I can go to school tomorrow."

"You go to school here?" I asked.

Biruté said, "Of course she does. She has to go to school."

"But you live in Kalimantan."

"She goes to school there, too. Wherever I go, Jane goes to school."

"Which one do you like best, Jane?"

"My school in Vancouver."

"Oh, but that's just because of your friends there, not because the school's better," Biruté put in.

Jane nodded. Biruté watched her dump the contents of several packets of NutraSweet into her tea and looked away. "How's your head?" she asked Binti.

Our meals began to arrive, and we tucked in to them in a very businesslike way. I tried to raise the topic of writing about her work, while nearby, on a TV set, a program caught Biruté's eye. "It's Karen Carpenter!" she said with a little flurry of interest, and the kids craned around for a look.

Jane said, "Who's she? I forget."

At this, a slight discussion commenced about our generation. Time, place, values. The two of us eyed each other sympathetically. "Look how skinny she is! It's absolutely horrible!" Biruté exclaimed. "Binti, look, that's what this society does to women! She had anorexia; she died of it."

Jane said, "What's anorexia?"

I said, "Really, I never thought about it being just this society. You mean women in Asia — Japan, for example — don't get anorexia?"

"In Japan they have other problems, but I don't think that's one of them. Anyway, it's not a problem in Indonesia. Men like women to be fat there. Well, not really fat, but big. Healthy. That's appealing to men."

The discussion dragged along. When was Karen Carpenter born? How old was she compared to Binti? Biruté? Me? We began to discuss our generation again, and I said that it seemed to me that despite many differences, we had much in common. We had married photographers. Gone off to islands. But Biruté had dedicated her life to Something, as very few women had. She had dedicated herself to an entire species. She had made a real difference. For this, she even gave up her own child. I glanced at Binti. To give up so much . . . I looked back at Biruté with all the interest I'd been gathering. And hope. Perhaps we would have dessert, coffee. Perhaps we would talk on into the night. I tried to put the scene in the Long Beach parking lot out of my mind, then considered the possibility that she was controlled by the OFI. Our dinner was not taking the shape of any social occasion I could remember, but I reminded myself that it wasn't a social occasion.

My guests were no longer hungry, but neither were they satisfied. They wanted to be home, and I could understand that. I could even empathize with them — another primate habit. I could see myself as Binti must have seen me, a nice-enough woman, one of a thousand who surrounded his mother, almost invisible except for the variation I brought to the restricted time he spent with her. To Jane, who reminded me painfully of my own daughters during the years they were dragged from one meeting to another so that I could keep my eyes on them and still earn a living, I was not even a distraction. To Jane, I was without a single, solitary justification for my presence in her life. I had not provided a stuffed rabbit, and any other possibilities were too far in the future to be interesting.

And Biruté? All week I'd been comparing myself to her when she was following an orangutan. I'd been trying to habituate her, exactly as she habituated them. Now I had her quietly assessing me from a branch overhead, but I had no idea what she was thinking.

"I knew we had something in common," I said, "when you mentioned Sharon Stone this afternoon."

"Why's that?"

"Well, my husband — the man I live with, I mean — is involved with a movie, and when I said she should play the lead . . ." (My suggestion had been based on one glance at a poster, but I didn't go into that.)

She blinked. "A movie? What about?"

"It's his book, not his movie. It's a movie of his book."

"Who's producing it? Could you give me the name?"

I said, "It's a novel." But she was paying attention for the first time.

"How did he get them interested?"

"Well, it's a beautiful book and . . . there was a prize . . ."

"What prize? Would my book be eligible?"

The waiter cleared away our plates. I had taken things in the wrong direction. I felt more and more fraudulent.

It wasn't easy squeezing into the car around all our packages, but it felt familiar. How many shopping trips had I made with how many women and children surrounded by how many bags of just this sort? It seemed to me that the best part of my life had been spent climbing into small, stuffed cars with tired children. "Has your mother been watching the trial?" I asked Biruté, as we rounded the corner and parked in front of the house.

"What trial?"

"O.J. They're almost neighbors. It must be pretty interesting for her."

"Is it? I don't know. I haven't really . . ."

"Well, I didn't either until I started watching down here. Yesterday I listened to three hours of testimony about the dog. It was incredible."

Biruté was opening the door. "What dog?"

"The one who . . . Nicole's dog! It's an Akita. That she got to protect her. The dog that led the . . . there were several testimonies from people who met the dog and just didn't understand what it was trying . . . the dog was absolutely trying to communicate, but everybody misunderstood, even though one guy really tried to figure it out and took the dog seriously and . . ."

"I didn't know about a dog. Nicole had a dog?"

"Yes! In fact it was the dog who found the bodies!"

"Found the bodies? A dog?"

"Yes, because when somebody finally figured out what the dog was trying to . . ."

The Geo was parked against Mrs. Galdikas's curb. Jane was asleep in a corner of the backseat, and Binti, I suppose, was tuned into the music of his own stars. But Biruté was finally home, and she wasn't leaving my car. For fifteen or twenty minutes we sat out there in front of her mother's house talking about an animal who had gone up a nearby street, trying every driveway, barking and howling at every stranger. "I wonder where he is now," I said, because the thought of the dog had been haunting me.

"You think he was trying to communicate with these people?"

"Absolutely."

"I didn't know about the dog." She looked out at the California night and its mysteries. "Well," she said, "I should go in . . ."

"I really would like to come out to Borneo." My voice in the darkness. "I really would. I could study with you there. Work on a book. Maybe bring my two daughters. One of them is teaching at a university. The other one is studying social work. They'd be great additions. I mean, if we want to learn about our relationship to nature, mothering comes into it. Nurturing, at least. I mean, you've adopted baby orangutans and raised them. Are we the only species that does that? And if so, why? You're close to something that might be the answer to so much of what I want to know. How we can readapt ourselves to fit back in to the natural world. This is so important and you seem to have managed."

She was staring past me again.

"We could do a kind of tutorial thing," I tried desperately, knowing that she often works with volunteers in that way. "I'd pay you. As if it were a class. I have money to do that. Money's not a problem." I seemed to be grabbing at thin strands of nothing.

"You have money?"

"I don't have money, but I have money for this."

Biruté pushed at the door and climbed out. I got out, too, and began unloading her packages. Binti got out and stretched and took an armful of them. "G'night, Binti."

"Night."

"Can I help you with those?" I looked at Biruté. If I were a dog I'd have had my stomach exposed. I'd have been on my back on her sidewalk.

"No, we're okay. Come on, Jane. But . . . well, listen, maybe we could just do a team of one. Maybe. Why not? I'll think about it. I'm going to London for two weeks but when I get back, I'll contact you. We'll see."

Large palms and tropical bushes brushed against a smudgy, starless sky. I felt like kissing the ground upon which Nicole's poor dog had walked. Everything was redeemed, including my faith. I could imagine myself in Borneo after all, with birds calling and orchids blossoming. Going up the river. Scent of things living and dead. I'd take a camera and a notebook and all the goodwill in the world. Biruté calls the rainforest Eden. She's gone back into the garden and made her peace with it. At Camp Leakey I would step out of a boat and onto a dock, and she would be waiting under the trees. She would stride along forest trails, casting her eyes up at the slightest sound and keeping a few voice lengths ahead of me. But there would be no need for conversation in the rainforest. We'd be listening for the overhead snap of a branch. Finally, when we had walked for hours, Biruté would stop, smile, point up and up and up, and in the distance, I would just make out, at the end of her gaze, a wild red ape swinging through the canopy.

How to Get There from Here

What happened between the last strata of the Pliocene age, in which man is absent, and the next, in which the geologist is dumbfounded to find the first chipped flints? And what is the true measure of this leap?

— Pierre Teilhard de Chardin

I went back to reading and waited for Biruté to call. I wrote her another letter. I made dates with the women I'd met who had traveled to Tanjung Puting, throwing questions at them about clothes and equipment. "Cotton, completely covered." "There's a new kind of pants at Mountain Co-op that dry in minutes." "Ziploc bags. Everything, all your paper stuff gets wet."

"Get hold of Carey Yeager," advised Anne Russon, the York University professor. "She's up the river studying proboscis monkeys, but she knows everything about the place. She used to work with Biruté. When she was doing her Ph.D." I made a date with a doctor who specializes in tropical medicine. We would need shots and malaria medication. I called the Indonesian consulate. We would need to renew our passports. From my perch by the phone I looked out at the melancholy weather. Snow, sleet, rain. My neighborhood in Toronto is a mix of derelict houses and houses of the same vintage that have been fixed up by enterprising people. I guess it's a neighborhood of both enterprising and not-very-enterprising people. There are men who sit on porches and men who don't have porches to sit on. They sit in the alleys that run behind the houses and their faces become familiar. In summer, they act as sentinels, but in March of 1995, their presence behind my house seemed ominous. Businesses were closing. There were no jobs. They had no place to sleep. And it was cold. While I was calling travel agents and consulates, I kept my eyes on a man who was mostly residing behind my house. He talked to himself. "I'm not so bad," he'd say. "I'm not a bad person." Sometimes he talked to my dog through the fence. "Don't worry, I like animals, I do. I'm a good person."

"How are you today?" I asked him one morning, as I went out the back gate.

"Bitter," he answered. And once he shouted, "Stay home!" as I went down our alley through the rain.

Leakey's study of the great apes was largely directed at research into behavior as it relates to skeletal structure and habitat. Tired of basing all his theories about human prehistory on the fossilized remnants of sapiens and presapiens who had lived thousands if not millions of years ago, he wanted to study people living much as he imagined the earliest humans had lived — modern hunter-gatherers — as well as our closest relatives, the other great apes. "Scientists were at last beginning to believe Charles Darwin's prophecy that the birthplace of both man and the great apes would be discovered in Africa," he wrote in *By the Evidence*. The idea that all hominids evolved out of the warm forests of Africa was radical, but eventually East Africa yielded remains of *Hominidae*, the taxonomic group that includes humans and apes, dated at twenty million years ago, when there were at least ten different species of apes living in the forest. In Olduvai Gorge in Tanzania and Koobi Fora in Kenya, two very different species of hominoid lived as neighbors for over a million years.

Finding the ever-narrowing gap between ape and man depended on recognizing the first use of tools, the first attempt at speech, the first time, even, that a tool was used to fashion a tool. The Leakeys — Louis and Mary — studied chipped pebbles as well as fossils. And because the fossilized bones were surrounded by fossilized seeds, insects, and plants, they were able to reconstruct the environment and to see how similar it was to the forested shores of Lake Tanganyika, which was inhabited by chimpanzees, whose habits might therefore provide clues to our own earliest history.

Leakey sent Jane Goodall out to study them.

Later he sent Dian Fossey to study the mountain gorillas.

Finally, he sent Birutė to Kalimantan, where an elusive tree-dwelling ape lives among the last of the human hunter-gatherers.

I bought guidebooks and maps, and when Kristin arrived for a visit from Portland, we spread everything out on the floor. An orangutan carries a mental map of all the trees in her range, including the state of ripeness of the fruits on each, but it must look quite different from the maps we studied. An adult male orangutan weighs around 180 pounds (80 kilos) and has four times the strength of a human being. The problem is how to sustain that huge body on a diet of fruit and leaves, so along with the map goes a sense of timing that is absolutely accurate. If he arrives at a tree too early, the fruit will be inedible, and quicker animals will plunder it if he arrives too late. So orangutan mapping skills require a complex system of neural transmission and memory storage. But maps are anathema to me. I can't seem to associate the lines on a piece of paper with three-dimensional reality, and I have no sense of direction stored anywhere in my brain.

Borneo was certainly far away. Somewhere between Malaysia and the Philippines, it was big — the third biggest island in the world — and had pieces of three countries in it: Malaysia (the parts called Sarawak and Sabah), the Sultanate of Brunei, and the Indonesian part where we were going, which is called Kalimantan.

"We can't get to Borneo from Bali. Not by air," Esta said on the phone from Guelph, after talking to a travel agent.

"So we'll take a bus to Java," Kristin decided. "Then we can get a plane."

"Birutė still hasn't written," I told my new friends, the former students and volunteers who now began priming me with stories of dangers and hazards, even as they told me what clothes to pack.

"She's testing you," they said.

"Loyalty," said Anne Russon, "is her big thing."

One of my new advisers told me how things had gone during her trip. They had begun their indoctrination in Birutė's village and then traveled upriver to Camp Leakey. But Birutė had been asked by the chief of police to ride with him, so they had not actually traveled together. "My initial impression of Birutė was of a woman with a will of steel," this volunteer said, relating how, as the river narrowed, the speedboat carrying Birutė and the police chief whisked around the volunteers' slower boat (*kelotok*), spraying water on everybody.

Another volunteer described a walk through the forest, during which she became dehydrated and vomited several times, falling farther and farther behind the group, while Birutė urged everyone to move along. Later, as the volunteer sat slumped against a tree, Biruté pulled out a large canteen and drank from it, saying it was important to prepare properly for a forest trek.

I dreamed that we were already there, that we were waiting at Camp Leakey for Biruté to arrive. Would she recognize me? When she finally came, she had many children at her side, but she had left me in charge of her smallest baby, and I stood there with the infant in my arms as she entered the room. The baby began to stir and cry and would not be stilled. She cried until Biruté came across the floor and took her out of my arms. But the moment was brief. Called to another part of the room, she handed the baby back, and I stood through the dream clutching a child I could not comfort, a child who wanted only her own mother. I was insufficient.

If the idea of traveling to the rainforest was new, my dreams of children were not. At night, my own are perpetually small. I am driving them through storms, up volcanic mountainsides, or I'm pulling one of them out of the sea. It's Kristin, my youngest, who always needs to be saved. She's having a birthday and I've forgotten the cake. I'm too busy or confused to do anything. Esta is the one who intervenes to save her, producing a cake and inviting the neighbors to a party.

Then one day in June, when Esta had finished marking final exams, she and I flew out of Toronto in the morning and arrived in Hawaii in the middle of the night, joining Kristin there for the first time since we'd all left together on a plane thirteen years before. I remember Kristin crying then, her head down on the tray table all the way across the ocean. They were leaving everything they loved because I'd met someone I loved. I was moving them to Canada to live with him. It seemed, even to me, entirely unfair, as if I were spiriting them to an unknown and forbidding piece of forest.

Now, Kristin and her father met our plane with 3:00 A.M. flowers, and within moments of entering the unforgettable humidity of our past, I was swept away to the small house where the father of my children lives on a hillside in Kaimuki. The house reminds me of our first one in the islands because of its painted redwood walls and the smell of earth coming up from outside and the cockroach in the laundry basket and the sense of this man I once loved now so near and so far away. But the words "once loved" make no sense. How is it possible to stop such a force? That night Kristin and I slept in his bed and I lay at window level examining the leaves of a mango tree as I used to lie in my own bed on this same island after he left us, hearing the first rhythms of my own language become words I could put to a page.

Once again we were being parents of a sort — two parents under one roof for the first time in twenty-four years — and I wondered what was left of the people we had been then. ("Where shall we take the kids tonight, honey?" he had joked in the car.) I couldn't sleep, but it wasn't jet lag. It was something else. Excitement. Nostalgia. There was even a small seductiveness to this.

Where is the life that late I led?

We were on our way to Borneo by way of Honolulu, Bali, and Java. We had avoided the costly OFI tour and assembled a rough itinerary, and we were excited and stunned. In the morning we stood together in the bright sunshine, blinking at the chatter of birds in flowering trees. The ex had gone off to church. He's become a Christian, very certain in his faith, whereas I am uncertain in all things. Inside there were fat three-ring binders full of images of my children. And me. Young — younger — youngest. As if we had never dissolved as a family. In the old days I could never look in mirrors, but now I pore over pictures of the past, finding my own face and arms eloquent. Hands — they are the same! But different. They're part of a picture of someone who lives but no longer exists. I can see myself as I looked to his lens but not as I looked to his eyes. That will always be another of the mysteries.

Where is it now?
Utterly dead.

We can't reenter our youthful bodies, but otherwise here, on this island, time seems almost to have stopped. The great ocean is below, making our hill seem simple, only geography. The cars on the highway circle the island at the rim of the ocean, and the houses cling feebly to any land made available, as if they depend on claws. Swash of green watershed. Swash of human habitation. Swash of green again. Your eyes will be opened, says the serpent, and you will be like God.

All of this is strange, but strangest of all is the grown son who belongs to my ex and whom I last saw as a boy. This child is now the age we were when we met. Esta is older than I was when we parted. All that tragedy in lives so new. My foot in his jaw, being swallowed.

After a day of circumnavigating our island, of stopping for "plate lunch" in Haleiwa, swimming on the north shore and crossing the Pali from the windward side at dusk, we flew out of Honolulu at 4:00 A.M., the hour of beginning and end, amid blue-shirted surfers who pressed against us on the plane, big-boned, muscular, and feral. Esta stretched out on the floor to sleep. Kristin did crosswords. Nine-month sublet? *Womb*. Eye flirtatiously? *Ogle*. Care for, four letters? I woke to Disney's Mowgli on a movie screen — monkeys hoarding jewels and treasure — as if we'd already arrived in the forest.

In fact, arrival in Denpasar was so sleep-deprived, so disorienting, that I only remember a long taxi ride into Ubud on a narrow crowded highway covered in smoke. The smell of a different land. In Bali, each village is arranged along lines of spirit. In the center, Brahma the wise, with Vishnu for life and Shiva for death at either end. Ubud is several villages grown together, and the lines there are complex. Durga, the goddess of Death, has a temple in the Monkey Forest at one end of town, and our hotel, a place of bungalows and offerings of rice and flowers, was on the other side of this forest. As the consort of Shiva, Durga is not a goddess to pass in the night, but we knew nothing of this, as we knew nothing of anything yet.

That evening — our first in Indonesia — we walked through the Monkey Forest to watch part of the great Hindu epic, the Ramayana, on folding chairs in an open temple in the middle of town. Part One: "The Abduction of Sita." Rama, god and prince, was exiled to the forest, and Sita, his wife, went with him. The blackness of the night was a curtain drawn across their journey, a blackness in which, all over Indonesia on different islands and different stages, parts of the old epic were unfolding, as they do every night — the prince being saved, in the end, by the divine intervention of Hanuman, the white ape. Within minutes, I fell asleep as two worlds, day world and dream world, collided. I was on the other side of the planet, and the rain had enclosed us in a wall of sound that drowned out even the clanging gamelans.

Hinduism was once the official religion on most of these islands. With the rise of Islam, it survives in Bali and in bits and pieces on other islands, where people sit under rain-drenched roofs and watch the epic in its many forms and with its clearly defined oppositions. Evil on the left side of the stage and virtue on the right. I wondered if anything could be so clear in life.

It was a night of no moon, no stars. Like sleep-walkers, stumbling and blind, we walked back through the Monkey Forest to get to our dry beds and more personal dreams. The white ape was a clue to something, but the thick trees over us and around us were ambiguous, and somewhere in the darkness were the monkeys we had seen that afternoon, hugging their young and grabbing at tourists' bags in hopes of food. Descendants of Hanuman and acolytes for the ancient, inaccessible Durga, the monkeys must have been everywhere, but they were invisible. Nothing was visible, not even our hands. We had skirted the huge temple earlier, on our way into town, but now it was darker than the sky. I told myself that this was a first encounter with the primeval forest. With Durga, preserver, destroyer, mother and origin.

Unusual fear. Thick silence. The three of us holding hands but moving single file, following Kristin's voice. "There will be twelve stairs here," she whispered, although she had only traversed the path one time. "Watch out. Yes. Now a turn. That banyan will be . . ." Kristin trained for many years as a dancer, and perhaps she remembers space through her feet. I hadn't noticed stairs or the length of the temple wall, but suddenly my youngest child — the one I dream of needing to protect — had taken the lead.

In order to get to an airport where there were flights to Borneo, the next part of our journey entailed a seventeen-hour bus ride into Java, ending in a mystic climb of Mount Bromo at dawn. Esta had read about the Mount Bromo experience, something that was going to take place on horseback, and "Wonderful!" she had breathed. So we signed ourselves up. The trip was a mere twenty dollars, including travel from Ubud to Semarang, which might have caused us to be suspicious, but we weren't. Seven hours, we were told. Your own jeep. Horses at the top. We did not even bother to pack lunch.

There was a procession of worshippers across the street as we climbed aboard the bus in Ubud. From the open window, I could see women in sarongs carrying baskets of fruit and other offerings on their heads. They poured out of an old stone temple and onto the sidewalk as if they were meant to guarantee the success of our journey, for like Rama, we were really moving into unknown space. Our hotel in Ubud had been recommended by Toronto friends, and we had had several sympathetic encounters there, including one with a young man named Hans Iluk, who knew all about Biruté and the orangutans and had offered much helpful advice.

Now there was a short ride on a ferry across the Java Strait; there was the issue of where do we pee and when do we eat. Those were the things we had on our minds during seventeen hours, bolt upright.

Java is different. Even in the middle of the night, it's crowded, male, and Muslim. As a woman, one feels no vibration except of the sexual sort, which is not sympathetic but prevalent. The funny thing about this vibration is that it's both curious and censorious. Hello, I love you, why are your arms so bare? It's not a good idea to be a woman, but to be a Western woman is worse. We offend with every breath and gesture.

Still, there was something miraculous about winding up Mount Bromo before dawn. "Where are the horses?" I asked, upon seeing our fellow travelers boarding another bus.

"At the top, Madame."

"And where is the jeep mentioned in the brochure?"

"What jeep, Madame? What brochure?"

"The one in Ubud. The one we were shown at the travel office where we bought our tickets for this tour."

"That is there, not here."

I tried to convey the image I had of myself driving a jeep up the picturesque mountain at dawn. I said, "I paid for a jeep," with my usual uncertainty.

"Then you can have one. Madame."

As our fellow travelers lurched away in their bus, we found ourselves being shoved into the back of a very small jeep. But I was triumphant.

In a few minutes I was going to mount the smallest steed in the world to cross the high plains around the crater of Mount Bromo in pitch darkness. Boys carrying tiny lights lit the way and anyone with any sense — including my daughters — walked behind them through the perfect, pristine dust. But *I* ascended at a little trot on a tiny horse. It was magnificent.

Bromo is an active volcano over 6,500 feet high with a narrow stairway of 250 steps up one side, which we travelers ascended like angels from all corners of the earth. And at the top we sat down together in the dark and waited for the sun to rise. We were very quiet. Grateful for the company of other humans, we moved about very carefully when we moved at all and looked down the back side of our narrow ledge into a crater that was smoking and sulfurous and had swallowed a visitor in circumstances just like ours only weeks before. Then we turned toward Mecca.

I suppose every person there was on some sort of pilgrimage. If I considered my own, it seemed unspiritual in the extreme. I was following a woman, trying to understand the effects of her work, what the cost had been and whether she had managed, even in a small way, to leave a trail of meaning. Had we been other kinds of apes up there on Mount Bromo, instead of *Homo sapiens*, we might have been watching the sunrise, but we would not have been thinking of Mecca. At least I didn't think so. What would we have been thinking?

I do not remember leaving the mountain or entering Yogyakarta later that day, although I remember the place we stayed when we got there. In Yogyakarta, we walked through a family's kitchen to get to our rooms, which were on a narrow street full of roosters and restaurants. I remember eating in a place where baby chickens ran over our feet and going by *becak,* a rickshaw taxi, to see the water palace. Like the Hanging Gardens of Babylon, the palace, Taman Sari, once had lighted underwater corridors, subterranean mosques, gamelan towers, and bathing princesses. Now, there was an owl tied by its feet to an iron ring in an underground room and, nearby, a bowl for coins. Human droppings. His eyes stared at us as we passed.

The great pools where the sultan's harem had once swum were dry as bone, but a guide explained how a woman would be chosen for an afternoon, picked out from the bathers and brought inside to a cool brick room. We were shown the brick room and the brick bed upon which the woman and the sultan lay. There is still a descendant living on the premises. Unlike other Indonesian sultans, Yogyakarta's has retained some of the ancestral power because of his father's loyalty to Sukarno during the war of independence against the Dutch.

One of the odd things about this palace — at least the modern parts — is that photographs of its great hall look the same whether right side up or upside down. I can't explain this, but I swear it's true. There is such a glossiness about the ceiling, such a glossiness about the marble floor. There are pillars that extend from one to the other. And not much else. The past is more real than the present.

In Java it was like that: we wandered, we posed for each other's photographs and for the photographs of everyone else. We carried our cameras from one angle to another, looking for the perfect shot of history, and there were thousands of tourists just like us or different. They shouted, "Mrs., take my picture, please" or "My picture with you, Miss?" until, amazed at the pleasure I could so easily provide, I began to fake the little clicks of a shutter being released. I said, "Smile," and pretended to press, because what difference did it make? "May I take your picture?" became an excuse to smile and nod and hold each other close. "May I take your picture?" made us aware of ourselves. In Java we became outsiders, but our fragile triangle tightened, exactly as I had foreseen.

I was finding myself astonished by my daughters' abilities, and it was delightful to watch them interact. One will bargain over the smallest purchase until the cows come home. The other spends on impulse, with no thought of consequence. Both feel the morality of their convictions around money. It isn't important, but it is. Or it's important, but it isn't. It had been years since we'd been alone together for more than a day or two. Now we were sharing everything:

every joy, every complaint, every decision. Even in the great expanses of Borobudur and Parambanan, we had nary an hour alone or apart. Only once or twice did we have separate rooms, and at those times I was reminded that I was the parent, that they were sisters, but otherwise we became, again, a single unit, functioning like a woman with six arms and six legs.

Finally, we went by bus to Semarang, where we could give ourselves one luxurious night in a hotel of several stories with telephones and hot water, before catching our Borneo plane. In the bar, someone sidled up to us and asked if we were with the OFI. It seemed incredible. "How could they be here?" I whispered, staring at my drink.

Esta said, "Don't be silly. Probably just a conference. Very common letters of the alphabet," and we retired to a room that was cool and a toilet that flushed and upon which one could sit. But when we unloaded our purses on the bed to examine our tickets, it seemed that Esta was scheduled to fly a day later than Kristin and me, which would throw the boat schedule off and leave us in separate disarray. So much for the luxurious room. Since there was a travel office in the hotel, we went back downstairs and spent the next two or three hours untangling our tickets with Air Garuda and the interisland plane.

In the travel office, a stranger came to our aid. "Can I help? I'm sort of used to this . . ." He turned out to be Eric, the very person I'd spoken to by phone at Bolder Adventures. So the OFI was the Orangutan Foundation International after all and we had all come, by some quirk of coincidence or fate, to the same hotel on the same day. The likelihood of this was just short of nonexistent, but it had happened, and we were scheduled to take the same plane to Borneo the next morning! I wondered if I had wanted this in some corner of my unconscious. I'd known the dates of the OFI trip and forgotten them. But whatever my unconscious may have felt, my conscious wasn't pleased. We were surrounded by overstuffed backpacks and egos. The atmosphere in the hotel restaurant swarmed with self-importance. Uneasily I began to see our trip through the eyes around us — as something frivolous and lighthearted.

The next morning, while the OFI group boarded an airport bus, we leaped into a taxi, speeding past them to get Esta into the ticket line first. Some of their tickets were also mixed up, but what hope had we, otherwise, against their tour guides and group influence? By the time they arrived at the Semarang airport, the smell of competition between us and them was already strong in the air.

Eventually all of us got on the same plane and flew over the same sea: the twenty or so OFIers at the back of the plane wearing shorts, baseball caps, T-shirts, and running shoes; three Spaldings in the middle; and another group (who were they?) up front. Under us, there was the shadow of a plane on the water, a plane that now seemed to be going beyond the beyond. Next to us, crouched in the aisle of the plane, was a woman named Giovanna who was too excited to stay in her seat. She was a teacher and had spent all her savings to come on the OFI trip and I liked her short gray hair and intelligent face. "We're going to stay at Camp Leakey," she confided. "And work with Dr. Galdikas."

The OFI material I had originally been sent described Biruté's research at Tanjung Puting as the longest, most detailed uninterrupted study of wild orangutans undertaken anywhere. "With the exception of four months, monthly observations have been continuous over nineteen years," it said. And in Biruté's words, apparently written for the volunteers the year before: "over the past twenty-two years I have systematically collected data on all aspects of orangutan behavior such as diet, foraging and ranging patterns, social organization, parturition, infant development, mother-offspring relations, reproductive behavior, vocalizations and tool use." It was easy to understand why Giovanna might expect to see her. But I was sure she was mistaken. "Biruté's teaching in Canada," I almost said, but something silenced me. The two of us were the oldest people on the plane and I felt drawn to Giovanna. Drawn because age was only one of a thousand things we had overcome to get here.

Remember there will be insects, leeches, high hu-

midity, malfunctioning machinery, fungus, toxic plants, creeping eruptions, malaria and who knows what else, the OFI material had warned, but we all peered out of the little plane's windows eagerly, as if what was in store for us was indeed Birutė's Eden. At last we were approaching that great thrill of green clinging by its roots to the island ahead of us, green that led away in all directions, borderless, astonishing, inarticulate. *Remember we are NOT in North America; we are in Borneo, one of the last wild places left on earth today.*

Blue Guides

In Pangkalan Bun, the OFI group was met by Biruté's friend and assistant, Charlotte Grimm. "Someone just as interesting as I," Biruté had said, so I looked her over. Like Biruté, Charlotte had married a good-looking, younger Dayak and had two children by him. Her husband and Biruté's are nephew and uncle, so she sometimes introduces herself as Biruté's niece. But Charlotte came to Tanjung Puting for the first time in 1984, with her wealthy mother, on an Earthwatch expedition, and when Biruté paid special attention to them by taking them to lunch, Charlotte announced that she would be coming back. To stay. Now they live in the same village, although Charlotte spends part of each year on the big island of Hawaii, where she is trying to establish an orangutan sanctuary.

The other group disembarking from our plane was English, younger than the OFI crowd and hardier, with long legs, clean faces, hiking boots, and rucksacks. They were volunteers for an organization called Trekforce, and they'd come to work for Carey Yeager, the proboscis monkey researcher I'd been advised to contact, whose research site is on the Sekonyer River above Camp

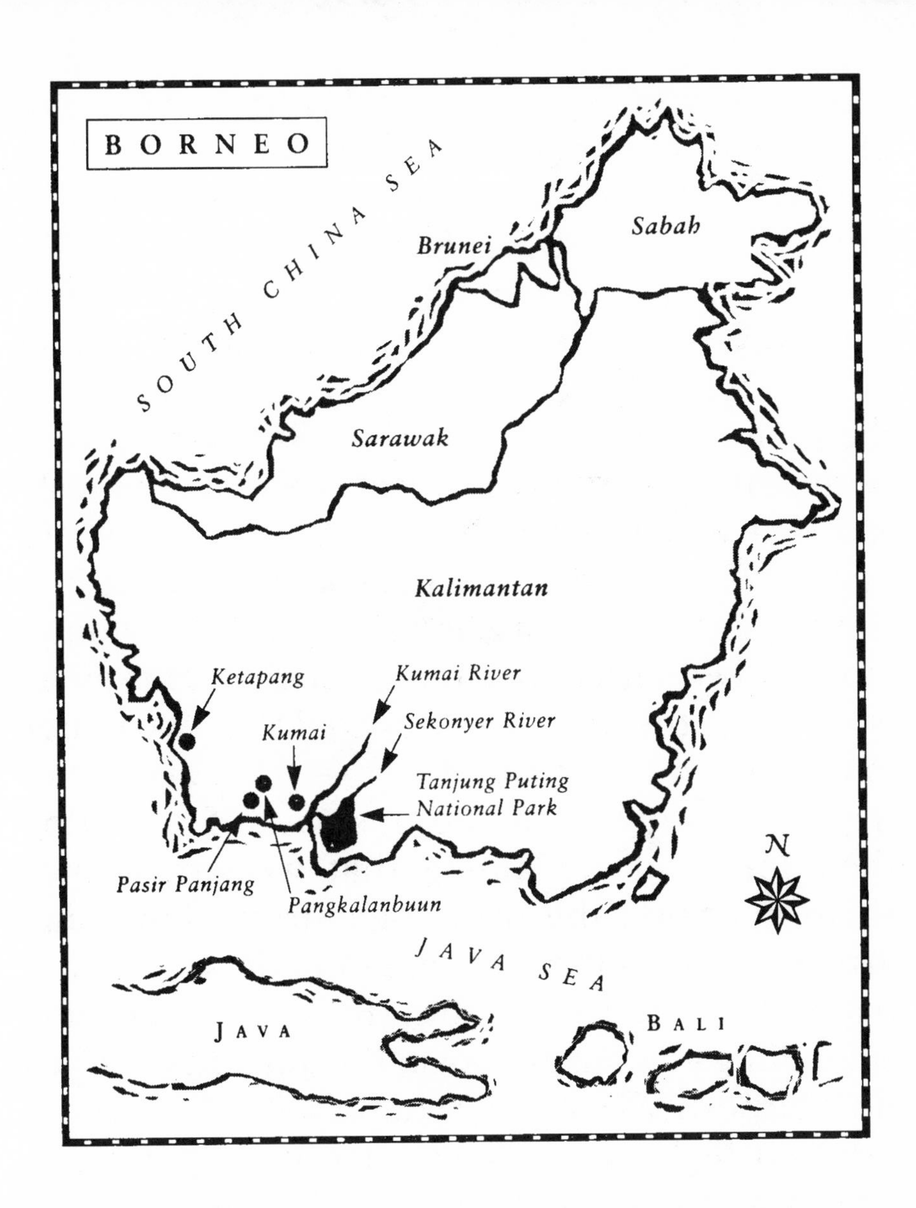

Leakey. Somehow the Trekkers reminded me of myself at seventeen traveling to Germany to dig postwar basements with the Quakers. Another set of footsteps in the forest, another set of good intentions. They were met by someone from Yeager's station, who also brought a letter for me. Since I'd had no word at all from Biruté, this small reception, with its apparent hospitality, was a comfort. In fact, since it seemed to confer official status on my arrival, it was a great relief.

There were other people waiting for us, too: a driver from the Blue Kecubung Hotel (with more mail and a car); a guide named Suwanto, who was connected to the hotel; and a young woman who had been contacted by Hans Iluk, the friend we had made in Bali. She was dressed in the uniform of a guide, but she was obviously too delicate, too female, to take us anywhere upriver. I glanced at her tied-back hair, her city shoes, and smiled at her shyness. The young man looked altogether more reliable. But as he went to get the car, we discovered that two of our packs were missing, and Riska — for that was her name — went about the business of reporting their loss and explaining that we would pick them up when the next flight from Semarang arrived. The young man was outside waiting impatiently at the curb, but Riska was exuding something pleasantly cool in the hot and crowded little airport. When she went off to find out about the next flight, we followed her as if by instinct, and Suwanto and the hotel car disappeared.

The Blue Kecubung was not what we had expected from the guide books, but nothing was. *Expect unavoidable, seemingly unreasonable delays, particularly in Pangkalan Bun and Kumai,* Birutė's OFI brochure had warned. Pangkalan Bun is a scrubby town on the southern coast. Two hundred years ago or so, the sultan of the old Kotawaringan kingdom moved his sultanate down from the banks of the Lamandau River to the banks of the Arut. The local people converted to Islam and became, like most coast-dwellers on the island, enmeshed in the vivid activities of interisland trade. Pangkalan Bun itself means "landing stage/anchorage/depot for munitions," and although it is not as close to the sea as Kumai, the Arut river is wide and navigable. There is a market strung along the riverbank, and boats of various sizes ply up and down. They may well be carrying munitions. They may also be carrying black market orangutans. There is a steamy center with two-storied buildings. There are mosques and two or three banks.

The Blue Kecubung is owned by a lumber baron known as Pak Aju (Mr. Aju), who has used his plywood profits to build an edifice that seemed to go on and on — concrete hallways leading to gaping renovations and unfinished stairs. By the time we arrived, the OFI group had taken over the half-finished lobby. Excited and hungry, they were waiting to be taken down to the basement restaurant, so we chose the road and walked into the hot center of town to "our best restaurant. Or at least," Riska added with a grin, "our most expensive one."

Her irony caught me by surprise, and I took a closer look at our quiet but forthright companion. So far Riska had suggested nothing except lunch and a little help in preparing us for the river trip, but over our first bowls of *nasi goreng* I asked her to consider coming with us. "As our guide. Would you like to? Would your boss let you off?"

"Yes. I will ask him." An astonishing smile. Already I felt we could not make the journey without this laugh, this quick step.

I told her that I had a letter from Carey Yeager promising me a twelve o'clock appointment on Thursday and the best boat in Kumai.

"That will be the *Garuda II*," Riska said. "That's Mrs. Biruté's boat. I mean, the one she always uses."

"But first I want to see Camp Leakey," I said. "Where Mrs. Biruté works. And the orangutans."

Riska said there was a check-in required by the local police (passports and photocopies), and she offered to do the paperwork and check on our boat. But that night Suwanto, the guide we'd ignored at the airport, sat down at our table in the Blue Kecubung restaurant and began to regale us with tales of his talents, flexing his triceps over our *mie goreng* and making it clear that we'd be better off if we did our traveling with him. He's a guy, after all, seemed to be the idea, and he knows his way around. And Riska is . . . well. There was a gesture that reminded me of the male gestures on Java.

"Were you born here?" I asked, because Riska had told us she was Dayak, born in a longhouse on the Delang River in the interior, and that seemed to me an excellent recommendation for a guide to the forest.

"I am from Java."

That settled it.

While he talked, he eyed Esta, as if his appreciation of her would guarantee her support, although in fact she was more irritated by his chatter than I was. Kristin wasn't listening. She was wondering about the *mie goreng*. Did it contain any meat? She was making friends with a cat. She was negotiating with the kitchen about dessert.

Our room had two narrow beds and a mattress on the tiled floor. Running water, hot and cold; a Western toilet; a TV; and bedbugs. When Esta proved this by displaying bites and Kristin started to scratch, I got up, put on my clothes, and began proceedings at the front desk with a phrase book. *"Ada serangga di kamar saya."* There is an insect in my room.

Blank stares.

I tried the phrase again — "There is an insect in my room!" I rushed across the lobby and got on the couch, jumping up and down to demonstrate the jerky dance of a bedbug, snapping my jaws and prancing. Soon the hotel boys were racing to the phone.

Whom did they call in this emergency? The helpful guide, Suwanto, who was asleep in his own bed in his own house and who, roused by the hotel's call, came cheerfully back to the Blue Kecubung to find out how he could be of service.

"There is an insect in my room," I said, again,

and Suwanto very sensibly sent for a can of bug spray.

But I asked for new mattresses, mindful of the story of the princess and the pea and knowing we probably appeared to be prissy Westerners. Why did I want more mattresses when I already had three?

Eventually I took one of the boys by the hand, and together we found a closet where mattresses were stored, and then together we carried two of them through the halls. At one point I'm sure there were seven young men and five mattresses in our room, and I was repeating that there was an insect in my room until there was nothing to be done but make all our beds and lie on them.

With Riska, we took a taxi the next day to Pasir Panjang, where Biruté lives beside a hot, flat stretch of road. This was in order to pass the time while we waited for the police to give up our passports and in order to satisfy some compulsion I felt to see Biruté's house, even though she wasn't in it. I had written to her here; I had even faxed her. Now I pretended it was research, but I wasn't very proud of myself as we pulled past the iron gates. Then I blinked. From her talks in L.A., I had thought that the house would be among Dayak villagers, but Pasir Panjang is a town with streets and curbs and satellite dishes with their heads in the clouds, and even so, compared with anything around it, anything within miles, the house looks very grand, set back amid trees, as white as a mosque, with a red tile roof.

Stranger still is the wall surrounding the property — so white that it shines in the sun and so long that it looks like an example of perspective stretching into infinity. "I thought her husband was a farmer," I said, noting several women washing its many sealed windows.

"Not any more," Riska said.

"What is he, then?"

"A gambler." Riska giggled. "But don't worry, I am good friends with Mr. Jacki, who works here." She walked up the drive to the house and spoke to one of the washing-women. The rest of us sat in the car. Riska had told us that Pasir Panjang was established by the Dutch when a particular group of Dayaks refused to be pacified. Since most Dayaks live in the rainforest, this was surprising. But nothing in all of Indonesia surprised me as much as Biruté's well-appointed house. A few blocks away we were to see a slightly smaller version when we passed the residence of Charlotte Grimm, but now a man appeared, as Riska came back, and she stopped and spoke to him. "Not Mr. Jacki," she said, when she got back in the car. "This man was one of Mrs. Biruté's workers, but he had a breakdown in the jungle." She sighed. "It's very hard out there. Now he's a little off. He can't get a job. I hope Mrs. Biruté looks after him."

Kristin pulled at my arm. "Let's go. Put your camera away."

"Nobody's home?" I asked Riska.

"There are some Dayak women and girls

working, doing cooking and cleaning and taking care of the baby orangutans in the back of the house."

"Here? Orangutans? I thought they were kept in the park."

"Ummm. That's where they are supposed to be . . . really . . ." Riska paused and seemed to check my face. "It's against the law, but Mr. Jacki says they want to build a clinic. Maybe out here or maybe in Pangkalan Bun. But there aren't many trees around and the road is so close. It's not a good place for orangutans."

I took another look at the house, and a thousand questions came to mind, but instead of asking them I asked Riska to ask the driver to take us back to Pangkalan Bun for our passports, then straight to the office in Kumai, where we could get our permits to enter the national park, our groceries for the ten-day stay, and I hoped, our bearings.

Garuda

There were moments when one's past came back to one, as it will sometimes when you have not a moment to spare to yourself.

— Joseph Conrad

Kumai is a boat-building town and it sits right on the river of the same name near its mouth at the Java Sea. To get there from Pangkalan Bun, one drives across the land from the Arut River to the Kumai or one goes by boat around a pointy chunk of Borneo that sticks out into the Java Sea, dividing two bays. The boats built in this seafaring town are enormous vessels made entirely of ironwood. Ironwood, hardest and heaviest of woods, floating here in preparation for great ocean journeys. Ironwood, which takes two hundred years to mature and which was once so prolific in the forests around Kumai that the seafaring Bugis came over from Sulawesi to build their famous and beautiful sailing ships. Many of them stayed. The ships are visible in several stages of completion in the water and on the riverbank, but before we got there we had still to walk through the hottest part of the day to the open market, which was shaded by blue plastic tarpaulins that cast a strange light over all of us, buyers and sellers.

Under those dark clouds I began to lose my connection to the others, who looked as unlikely in that place as the vegetables piled around them. But even my own hands and feet had begun to look strange. A group of little boys screamed when they saw us, as if we were monsters. The heat was an entirely different substance than it had been in Bali or Java. It was unbearable, oppressive and ovenlike. Solid shapes wavered, melted. I was faint. We needed potatoes, cabbage, tomatoes, bananas, rice, water, canned fish. Riska kept urging me to walk, to make decisions, to count out my money and pay for our supplies, while the little boys, at our slightest move, became frightened and ran into the racks of clothes their mothers were selling. I thought of Birutė's white house and understood the sealed windows. Air conditioning. On the point of tears, I knew I would never last out the trip. I was too old. I wasn't fit. I should have torn off all my clothes and leaped about to underline this: *Send me home,* I should have shouted, wanting to lie down in the filthy marketplace and quit. *I hate all this. I can't breathe.* The clouds were descending, as they sometimes descend. I had gone past my limit.

The people who really needed me were on the other side of the world. Michael, my mother, stepchildren, old dog, old cat. What was the forest to me? I had no role to play in it and no energy to spare. There would be no phones where we were going. What if someone at home had an emergency? Since my brother's death, I am my mother's only child. I knew she would worry constantly, and now the whole trip seemed utterly selfish. My children would hate the trip. We would learn nothing useful or interesting. The boat we had rented would carry us away from everything we understood, everything we collectively believed. The last fine thread of our past would break.

It did not occur to me then that Biruté must have felt this in much larger doses as she'd set off, twenty-seven years before, for the remote, unbuttoned wildness of Tanjung Puting. For months she lived among strangers, or no one, with only her husband for familiarity, having left everything she knew and given up everything she had in order to learn what she could about orangutans. It did not occur to me because I wasn't thinking about Biruté anymore. I wanted a fax machine. A telephone. Contact. I should have called Michael in Pangkalan Bun while I had the chance; now it was too late. I missed him. He might be worried. My nonchalance in the face of what was ahead had been stupid and prideful. Now I had to sit down, wipe my eyes and face. I wanted not here and now, but future and past. My own, not this.

Around me were people who saw me only as *outsider*. And how did I see them? Kumai's population is made up mostly of inland people who long ago moved to the coast. They are called

Melayu. To *turun melayu* is to come downriver from inland Borneo. To me — born in Kansas of plain old European descent — they were exotic, partly descended from the Negrito Australoids who arrived in this part of the world forty thousand years ago, who swept out of Asia and into the shade of the great, already ancient dipterocarp trees, and partly from the Mongol Austranesians who came later. In the middle of our century, Tom Harrisson found a 35,000-year-old skull in the Niah caves of Sarawak — the oldest *Homo sapiens* skull ever found — but when it was the brain case for a living, breathing man, he had already forced his Negrito cousins into the hills after taking a daughter or two back to his cave. His descendants are upriver Dayak like Riska's relatives and the coastal Muslim Melayu in the Kumai marketplace, where I languished, miserably hot, and began to notice their faces: an old gentleman who squatted at eye level among his potatoes, a woman so beautiful that I asked Riska to ask her if I might take her photograph.

Riska said, "She doesn't mind."

"Are you sure?"

"She likes having her picture taken. Everyone wants to take it."

Deflated, I took the photograph. But looking through the lens was somehow a means of taking back my point of view. I began to feel better. If nothing else, even without language, I could see. And Borneo as I framed it would be what friends at home would see, so I had to make choices. I could choose one face over another — and perhaps I could even select enough potatoes for our boat trip.

With a ten-day supply of food, we made our way to the offices of Perlindungan Hutan dan Pelestarian Alam (PHPA), the Ministry of Forestry's Nature Conservation and Forest Protection Agency. It stood at riverside, and we paid a secretary a small tip to hurry to the typewriter and compose our park permits. "Look outside," Riska said, nudging my elbow as I stared at a hand-painted map on the wall. "Look, it's the *Garuda.*"

Excitedly, we all ran to the door and saw at the end of the dock what must be the bluest and brightest riverboat in the world. Riska waved. "It's Yadi," she said. "He's the best driver on the river. This is Biruté's boat — the one she always reserves — and he's her driver. He's always loyal to her."

Loyalty . . . I had heard weeks before . . . *it's her big thing.* We carried our packs and boxes to the dock, where Yadi, a young man of twenty-three, stood waiting to stow them on the lower deck of the two-tiered boat. Her driver . . . what would he think of us? It was a small worry, but I couldn't shake it.

Kelotoks are the traditional means of transportation on Borneo's rivers, but they vary in size and age and condition. The *Garuda II*, which was to be home and transport for the next

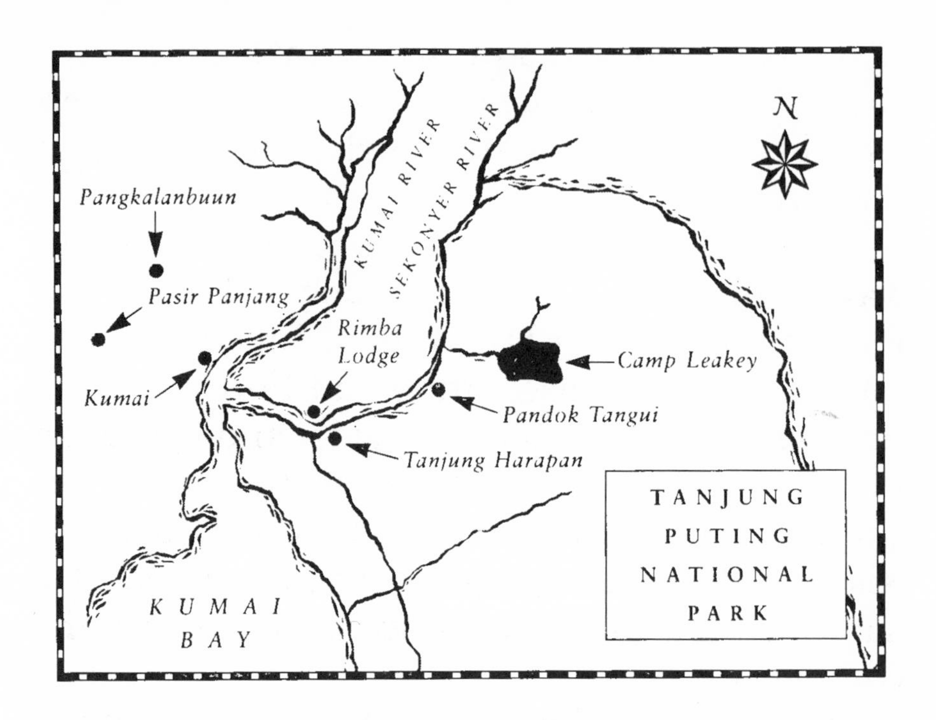

ten days, has a single engine, a seat on the lower deck for its driver, and a kerosene stove. Otherwise, it is bare as the sky, and the same color except for its trim, which is brilliant turquoise.

From the hand-painted wall map in the PHPA office, decorated with proboscis monkeys, orangutans, kingfishers, and giant hornbills, I had learned that the Kumai and Sekonyer rivers meet just south of town, where the expanse of water is so wide that our kelotok would seem minute. Now, engine puttering, we slid past the great ironwood ships, so raw and immense that they look prehistoric even as they are being built. Amazing that these great river hulks — shaped exactly like arks — can float, although one lay solemnly in the shallows as if expecting God's local creatures to march up the riverbank in pairs: two clouded leopards, a pair of macaques, a lone sun bear, then his reluctant mate.

The river takes half an hour to cross. Barges pass, a few boats, and the town quickly slips behind, out of sight. On the other side, angled through a thickness of trees, the Sekonyer was invisible until we actually entered it. Along its banks, great swaths of green connect water to land, and they rustled as the pressure of the *Garuda*'s breath forced them back, making the sound of stiff skirts and a thousand women leaning together. Patient. Impatient. Pulling back on either side of us like curtains as we entered the narrow river and took our first scent of the forest.

There was the rustle of pandanus, the cry of nipa palms, but we were loud enough to provoke silence in everything else. In this way we were intruders, with our bright boat and our noisy motor, although we soon began to feel the rhythms of the river and its reeds seep into us and to feel the new rhythms of ourselves as a unit. We were four women now. There were two men aboard, but they were them, not us. There was boyish Anang, who stayed close to the stern, and Yadi, at the wheel. Nevertheless, for purposes of conversation, decision, and society, we were four. Women. A circle or a square but not a triangle. A shape for which a new balance had to be found. How and how much to include Riska in our thoughts, our jokes, our conversations. When and how often to explain ourselves in order to be neither condescending nor exclusive.

Riska was born in a village about seven days upriver, and the water is a different kind of trail for her than it is for us. For her it moves backward, toward the place where her father was a hunter, her mother a teacher, toward the Dayak village of Kudangan, where she was born and spent the first seven years of her life. For Riska, this is the long road to her past and her history. The Dayak live along rivers, but depend on the water less than on the forest. Their animistic religion, Kaharingan, is so old, so aligned to the land and all its powers and secret places, that the Indonesian government has had to recognize it, although in order to put a good face on it, they have named it a branch of Hinduism, which is ridiculously inaccurate.

We were entering a landscape in which people have lived much the same way for thousands of years. Rattan for traps. Pandanus for baskets. All the necessities are guaranteed by the forest except for the rice grown in swidden — or "slash and burn" — plots, small sections of land temporarily stolen from the trees. On the river, narrow sampans skim the surface, nets and lines dipping, faces staring into the depths. Where river meets swamp, there are houses built of poles, around which swim prawns and spawning fish.

The creator is male and female, upperworld and underworld, heaven and earthly water, hornbill and serpent, but lesser deities control the mundane affairs of human beings, and there are spirits, both friendly and malicious, who live in the forests and rivers, in stones and animals and trees. Even today the Dayak are relatively untouched by urban ways. Scattered over a vast area of intractable swamp and forest, these tribes have had little contact with the government except through the village elementary schools and sporadic, understaffed medical clinics. This is why Riska's mother was not allowed to stop teaching. The village officials did not want to lose her and forbade her to leave. "Escape" is how Riska put it. But one dark night, with only a few belongings, the family boarded a sampan and paddled for seven days. The children were fastened into cages on the little boat so they wouldn't fall overboard, a fact that Riska remembers vividly because of her fear that the sampan would capsize on one of the many sets of rapids and she would be trapped underwater in her cage. "I remember the night we left the village," she told us. "I did not know what was happening; nobody knew. But I saw my mother crying."

After a frightening time of poverty and displacement, Riska's mother found a teaching job in Pangkalan Bun, but even now she comes home from school every afternoon to an unemployed husband and son. When Riska mentioned this, she managed to convey her frustration without saying anything negative about either parent. She was twenty-five and not married, too old, she said, to be attractive to an Indonesian man. Saying this, she braided her waist-length hair and covered it with a cotton hat, worn to keep the sun off her skin.

For miles and miles we traveled with only the palms at our sides, though they were taller than the boat and provided a perfect unassailable screen. More animal groups live in these swamps than in any other part of species-rich Borneo. Here, fish climb trees and spit at flying insects. Here, land is born as seedlings root themselves to silt. Silt becomes mud. Mud becomes clay. Clay becomes earth as it mixes with fallen trees. Earth becomes forest and forest becomes peat. The soils have a pH range of 3.8 to 5.0, and the dark, tea-colored water is anything but sweet. Tanjung Puting is a tiny part of the island, but it is big enough to be visible on any map. *Tanjung* means "peninsula," and its levees are being built even now by rivers running over their own banks and leaving finer and finer particles of sand as they flow toward the sea. Where the particles rest, small depressions are formed. Sometimes gold is laid down by the water. But it is only one of a thousand things.

Conveniently for anyone making Rousseau's twin journeys, in the Dayak world there are two souls. One lives in the body. One travels in the world of dreams. The *Garuda* was now transport for both, comforting in the harsh sunlight, when the upper deck created shade for anyone underneath, and comforting even in the rain, when two blue plastic wings were unrolled and let down over the open sides, making a dark cocoon in which we could easily fall asleep. In fact, the effect of the water, the forest, and all that wavy blue became hallucinatory, so that I sometimes didn't distinguish between fact and dream.

At times, we seemed to be in the *Ramayana* again, for Vishnu the preserver had ten incarnations, and Rama was only one of them. As Rama, though, he rode the wings of the sun-bird Garuda during his exile. It seemed strange that the great story of Asia involved exile *to* the forest, and in the Western tradition we were thrown *out* of it. Now, like Rama, for the next ten days we would ride the *Garuda*'s back. Sitting on top, in the open breeze, we spread our wet laundry out, making flat, reclining bodies of our clothes after washing them every afternoon. Below we slept, sat, stored our things. Riska made coffee, tea, and meals in the tiny galley, and we ate above or below, depending on the weather. (To lie on the green slats of that upper deck and watch the palms and sky slide by and an eagle sail between them was like . . . being a god with a pair of wings.) We would look for proboscis monkeys, the bulbous-nosed clowns who live along the riverbank in small tribes. We would look for kingfishers, and while we looked and talked and looked, our circle would change. I would be young or old. I would be one of four or a little alone, off to one side.

It was easy to slide into meditation. Aside

from our bodies and the food to feed them, the lower deck had a stack of thin folding mattresses, pillows, and our overstuffed packs. This meant that when we were below, we sat with our legs tangled and our bodies half reclined. We spent several hours each day like that, and all of us slept in the cabin at night like seeds in a floating pod, the *Garuda* conveying both souls, one through dark, one through light.

At the rear of the cabin there was a door about eighteen inches high, through which we crawled to get to the galley or to a cubicle that stuck out over the stern in which, under cover of four shoulder-high walls, it was possible to pee or shit or bathe with a bucket of water, since there was nothing between the flesh and the river but a few wooden slats. The galley was a small space at the stern with a gas burner on the floor and a couple of pots, not unlike most kitchens in Indonesia, except that it was blue and required some careful balancing on the part of the cook. The food we'd bought at the markets in Pangkalan Bun and Kumai was stored in the cabin where we slept and ate and did whatever we did, which wasn't much because we couldn't move, except at a crawl, and how is it that a baby does that so easily? After a day, my knees felt swollen, chafed. When I crawled over the various six-inch thresholds between cabin and kitchen and the topless cubicle on slats over the river, I longed for knee pads. In other ways, too, it was awkward, being so much older than the others. I seemed to pee more often than they did, and the sound of my peeing seemed to reverberate off the water like an announcement of my difference. I caught myself groaning as I heaved myself up the tiny ladder and crawled onto the upper deck. I was graceless in the tiny bath cubicle and too shy to ask Anang, who might be hovering just outside, to pass me the soap I'd forgotten to take in with me. Since baths in the thin-walled box came from a bucket that was lowered into the river at the end of a rope, they weren't baths at all, but showers, or what Indonesians call a *mandi.* The water was icy cold but the air was hot. At four in the afternoon these buckets of water redeemed everything.

Yadi's father is a Bugis with three boats and two sons. Boat-builders and navigators for two thousand years, Bugis regard a son as essential, and Yadi was expecting one of his own. At twenty-three, he was already self-assured. He spoke no English, so being with him, one was physically accompanied but left mentally alone. Anang remained invisible. Mysterious, it was, to be on a twenty-foot boat with a silent driver and an invisible hand.

The sound of a kelotok is a loud *puttputt,* a churning of the river, and it throws an echo against the trees and nipa palms along the bank. When we'd left the broad stretch of Kumai, we'd moved into the silence of the rainforest, and when the echo came back in a narrow channel the sound was deafening, as if we were at war, as if a helicopter were circling overhead. I kept thinking about Biruté in the early days. I imagined her first trip up this river. I thought about the tension and excitement she and Rod Brindamour must have felt and their separateness that day, which must have been more obvious than usual, as separateness always is when we come close to something important. I thought of Biruté young and determined. Age changes us, but character doesn't change.

As we moved up the Sekonyer and away from everything familiar, it began to rain, then to pour. Suddenly life, Riska, the four of us, this beautiful blue-and-green boat . . . how had it all come together? A great joy hit me, as if it were part of this most ancient of airs. We sat below, above, below. Upperworld and under. Man world and woman world. We lay with our mouths open and drank in the sky. We sat with our eyes open and took in the rain that fell on the blue plastic wings that had been dropped to keep us dry. We sipped hot tea and smelled the river. Time went by and never moved. Two hours or four or ten of fleeting, eternal sky.

Fruit Stares

Rousseau, through an admittedly wholly speculative reconstruction of man's state of nature, managed, in 1755 to produce the best available description of the true orangutan of Southeast Asia for more than two hundred years.

— Robert Wokler

When did we stop? We stopped eventually. Tanjung Harapan, the first station of the national park, is four or five hours from Kumai (though only about ten miles on the map), and there, right away, we saw our first orangutan, red-haired and oddly familiar, like a relative we'd surely met sometime, somewhere. Riska called her Melly at first glance, though, much later, weeks afterward, I was still astounded by this feat, for to me most female orangutans look much alike. Genus: *Pongo*. Species: *pygmaeus*. Orangutan is Malay for "man of the forest." They used to live all over Asia, but now they live only in parts of Borneo and Sumatra, divided for so long a time that the two groups are assumed to be separate species.

Somewhere on the boat I had left the instructions I'd received from the OFI, but Rule #3 had been easy enough to memorize — *Never initiate contact with an orangutan* — and it was repeated on a pamphlet we'd been given in Kumai. What to do, then, when Esta sat down on a log and Melly sat down next to her? It was raining and orangutans are supposed to hate getting wet, but Melly was out in the open near the dock, half human in size but more than half human in every other way. *Initiating.*

In fact, orangutans share 97 percent of our DNA, which means we have a common ancestor. Which means, in evolutionary terms, we split off only recently, about thirteen to eighteen million years ago. There are differences, of course. Our molars are smaller. Our jaw is weaker and we have no sagittal crest. Our canines do not project above the level of our other teeth, so we have a hollow in the cheekbone where the root of the longer canine used to be. There is the matter of brain capacity, too, but intelligence cannot be measured in cubic centimeters, as Charles Darwin once observed. Those, like Biruté, who have studied this red-haired ape believe its formidable brain has been shaped, to some degree, by human persecution. Other creatures do not look us so closely in the eye, and never with eyes so like our own.

Esta looked at Melly. Melly looked at her. The orangutan's stare is intrusive. Still, it is a stare without judgment, more like a photograph than a mirror. It is a stare developed over millions of years of looking for edibles in distant trees by bringing each particle of forest slowly into focus.

Because of this, Biruté calls it a "fruit stare."

The OFI group had arrived at this first station before us, and several of them were staring, too. Barely recognizable under plastic ponchos and raincoats, they seemed not very pleased to see us. We exchanged greetings and waded through mud and wet grass, ponchos and other rain protection making us hot and sweaty and awkward. One of the orangutans had gone under a cabin to get out of the rain, so we walked over and bent down and peered into the darkness, the way the Munchkins must have done when Dorothy's house fell on the witch. Expecting to see two legs sticking out and a pair of red shoes, I saw only mud and a few blades of trodden grass.

"Where'd he go?" we all said, and these were words we would hear ourselves saying many times in the next few days.

In the forest, orangutans build nests when they feel the rain coming, but these ex-captives have no nest-building skills. At least not yet. They were there in the park because they had been confiscated by the PHPA rangers or police. They'd been confiscated because it is illegal in Indonesia to hold an orangutan on private property. It's illegal because so many members of this endangered species are murdered in the process. Mothers especially. Some people calculate that for every baby kidnapped, two have died in transit and all three mothers have died defending their babies.

Alan, a ranger working for the PHPA, came

over to exchange news with Riska. "Where's Gistok?" she asked, then said to me, "The naughty one."

"In jail. Do you want to see him?"

Cascades of water blasted down from the gabled roof of the place he pointed to. The buildings at Tanjung Harapan are built Melayu-style on stilts with shallow verandas. We ducked under one of the waterfalls and up onto a wet porch to peer into another darkness — this time a locked room where a seven-year-old male orangutan sat clutching a yellow blanket and glowering. Gistok, in a former incarnation, belonged to a man who gave him birthday parties every year. Now the man visits on Gistok's birthday, and according to Riska, although he doesn't remember the owner, he likes the annual cake. I recalled a story about a gorilla named Koko who learned sign language. When Koko was shown a photograph of herself at her birthday party, she signed "me love happy Koko there." Perhaps Gistok, too, remembers his past. Perhaps he wants it back.

"What did he do wrong?"

"Broke into our kitchen and stole our rice."

As an ex-captive, Gistok has never lived at Camp Leakey, the third station in the park. This is a new policy for the park. No ex-captives are being introduced up there because it's been decided that orangutans who have lived with humans should not associate with wild orangutans. While this may happen anyplace in the national park, at Camp Leakey, where there are many ex-captives surrounded by many wild orangutans, Gistok would compete for food and space and bring a threat of human-borne disease. At Tanjung Harapan, once a human village, encounters with the wild are less likely and there are fewer ex-captives in camp. Perhaps that is just as well for poor Gistok, who was only trying to be a man instead of a "man of the forest."

When the OFI crowd retreated to the Rimba Lodge across the river from Tanjung Harapan, I decided that spending a night there would be part of my research. These people were interested in Biruté's work, too, after all, and probably knew more about it than I did. Maybe I could profit by a friendly chat or two. Maybe I was a little nervous about living in a boat on the water. Maybe I felt shy about sleeping with Riska and the boys. At any rate, Esta, Kristin, and I checked into one room (two beds and another mattress on the floor) and ate in the dining room, watching our airplane friend Giovanna and the others eating at their long, central table. Rimba rooms are expensive by local standards, but it's the required resting place for anyone on an OFI tour. Visitors no longer stay at Camp Leakey in the "team house"; the PHPA has made all four stations in the park off limits after 4:00 P.M. The lodge has a nice, big dining room built to hold the Great Ape Conference of 1991, and there were plenty of empty tables. That night, however, despite our shared flight across the Java Sea and our shared adventure with the orangutans across the river at Tanjung Harapan, none of the crowd at the central table spoke a word to us. We smiled at those who met our eyes, but there was no response. What was wrong? It was as if we weren't all visitors from another world. Charlotte had been joined by her children, nanny, and husband, Uil, but none of them spoke to us either. "I guess Miss Charlotte knows we are going to see Dr. Carey," Riska whispered.

I was getting used to hearing first names attached to titles, but the complexities of social relationships were something else. (Riska was eating with us, for example, but Yadi and Anang were still on the boat, and even when we all ate within its confines, they stayed in the galley.) When I asked Riska why seeing Dr. Carey was a problem, she said, "Mrs. Biruté is mad at them — at Dr. Carey and Mr. Trevor, her husband." I wondered why the OFI wouldn't be delighted to have another research station to visit with its tour groups, but Riska asserted that they aren't even told about Carey Yeager's work, in spite of the fact that she's an English-speaking scientist doing important animal and rainforest research, a scientist who worked with Biruté for two years at Camp Leakey. I wondered what could have happened between "Dr. Carey" and "Dr. Biruté" to cause such a rift, but Riska had no answer to that.

After bowls of corn chowder, we left the dining room and went out into the darkness. We should have stayed on the boat, I thought then, having known the comforts of an inside bed all my life and finding the Rimba Lodge less friendly than I had imagined. There were crocodiles in the dark Sekonyer water, but at least their defenses were comprehensible. As Riska left us to go to her mat on the *Garuda*'s blue deck, we chafed at the thought of our room and its door that would shut us off from the great, overwhelming night.

When I was a child, our neighborhood Girl Scout troop was run by a team of our own mothers, whose idea of a "camp-out" was clean bedding on someone's living-room floor. Vaguely, we girls knew that something more should be expected of us in order to earn the badges our mothers handed out so generously. Vaguely, even then, I longed to taste the night, and now, with it stretched between palm trees and mangrove swamps just beyond our wall, I was shut up again, inside, with no one to blame but myself.

I wasn't sure what I was after in the Rimba Lodge. It's nicely designed, with plenty of wood and decks that stretch between its wings and circle its rooms so that the swamp underneath can rise and fall as it has always done. There's a gibbon who hangs from a tree, and there are monkeys who run across the roof, and there is the lobby with some rattan furniture and the check-in desk. Our room was lit by a single bulb and cooled by a ceiling fan. We each took a shower under cold running water — our last for some time — and then shut off the light. There was the sound of voices on the other side of the wall, but they only made us feel more separate. After two weeks we were adjusted to the time zone, but now there was another zone to enter, and lulled by the frogs who sang us into unconsciousness, we began the kind of wandering that only takes place in sleep.

The next morning when we crossed the river to watch the first feeding at Tanjung Harapan, Gistok, the "naughty one," came bouncing down the slope between cabins and dock, turning somersaults over his prehensile feet. Gistok is about four feet tall, and most of his body is arms. For swinging through trees, he is wonderfully designed, but for walking, he has certain disadvantages, especially those feet, which toe-in awkwardly. This is why his wild relatives mostly stay aloft.

Watching his bumbling approach, I could not imagine what his intentions were. *If they attempt to contact you, retreat slowly and with dignity.* That's what was going through my mind. The damned brochure again. And Gistok looked bigger and bigger as he got close. *Orangutans are slow moving and usually don't run.* No, but then . . . he was in front of me — *If by any chance an orangutan does grab you and you are uncomfortable, stay calm; do not make any jerking movement* — and with a strong tug, he jerked me away from the others, as if they had nothing to do with us, as if my life, from now on, would be in his strong hands. I heard myself saying, "Yes, all right, I'm coming," but really I had no choice. Once this orangutan had me, he held on.

Keeping tight hold, pulling me forcibly across the sandy ground in front of the rangers' cabin and on, past another cabin, he suddenly stopped, my hand firmly in his. We had come to the last cabin on the grounds, the one closest to the forest, and as he bent over carefully to look underneath the floorboards and porch, I bent over, too, and stared into the dark, where one of Gistok's companions had been hiding from the rain the day before. What were we looking for now? Gistok straightened up. Pulled me along to the trees. Clearly, we were heading into the forest, though I didn't know then that Gistok had no more knowledge of that forest than I had and certainly less interest in being part of it. All I knew was that I was the captive of an ex-captive orangutan. My tone had changed. "Gistok, please!" But he was much stronger than I and much more determined. If I balked, he pulled, pushed, even nipped.

Where were my children? Riska? The station rangers? Why weren't they doing anything? There are stories of rape, violent attacks . . . but kidnapping?

I had begun, ever so slightly, to panic.

It seemed strange that all this had begun with Leakey and his interest in the great apes, whose physical and social environments, adaptations, and behavior he thought might shed light on the lives of the early hominids. Stranger still that an orangutan who had been "saved" by one of his protégées was pulling me back to the trees.

It seemed that I was the one who might need to be saved.

Then someone shouted "Gistok!" and I turned my head to see Alan, the station ranger, chasing us with a stick as Gistok dropped my hand and shambled off sheepishly to hide behind a bush.

These rangers are young men about Yadi's age who work for the PHPA. They're hired locally, usually from Kumai, although the supervisors in the office are professionals sent over from Java.

When Biruté and Rod first arrived, there was no PHPA in evidence. Tanjung Puting was a game reserve established in the 1930s for the protection of orangutans, and aside from the other humans living along the river's edge, they had the orangutans pretty much to themselves. In 1982, partly at Biruté's insistence, the reserve became a national park. Now its brochure, complete with maps and color photographs, announces that one of the aims of the park management is the protection of the many endangered species living there, an aim that includes the "re-habilitation programme whereby apes are confiscated from captivity and reintroduced back into the wild under close observation."

There is a wonderful phrase to describe this

process, one that would raise the hackles of fundamentalists and creationists. According to the PHPA brochure, the rehabilitation project is working on a principle of "progression to the wild."

After the feeding we continued upriver to the second station, Pondok Tanggui, where I was hoping to meet the young veterinarian from Java who works for the PHPA. "Be sure and meet Abdul Muin," Hans Iluk had told me on Bali. "He's worked with Herman Rijksen and Willie Smits, and he's very bright. He's part of the new way of doing things . . ."

Biruté has always cared for the ex-captives by feeding them milk, bananas, noodles, and cooked rice, the idea being to build up their strength, to get them healthy enough to survive in the forest. And the morning feed at the first station had not seemed very different from that, except for a lack of starch. A few young orangutans had come swinging down from the trees, some of them with infants clinging to their backs, a tiny head peering out from under an arm or behind a neck. Orangutans move through trees slowly and carefully, as if they are infirm, testing each branch to be sure it will hold them, a skill that takes both wit and practice. As the rangers are throwing down bananas and pouring milk, babies cling to a mother's head or shoulder hair and sometimes lean over the bowl and take a few sips. The older ones approach, drink, grab as many bananas as possible, and then beat a fast retreat. They peel and chew, peel and chew, and if they are carrying infants, they give them the pulp with their lips. The rangers occasionally cluck and chase off a greedy milk drinker, and they make sure that Gistok doesn't get more than his fair share. Of course this isn't "natural," but these mothers and babes aren't living in the wild. They're in training for it.

What was the new way? I wondered.

Back on the *Garuda*, back on the river, back under the sky, the morning opened up. A kingfisher flew overhead, one of the sky deities who have protected this island for thousands of years. The Iban of north Borneo say that during the great deluge each race of humans escaped with their written words. The Chinese carried theirs on their shoulders, the Europeans put theirs into their hats, but because the Iban carried theirs in pouches around their waists, the words got wet. When the Iban returned to land, the words were put out on rocks to dry, but a flock of birds carried them away, so the Iban have no alphabet. All the knowledge of their ancestors is obtained from the birds through messages.

While we chugged along, I thought about the conversation I'd had with Charlotte's husband, Uil, over a cup of coffee at the lodge. "I own part of this place," he'd said, looking around the dining room proudly. I'd said, "Really? Then . . ." but Charlotte had called him sternly away from me. "He'll be right back," she promised. But he hadn't been. In fact, those were the last words I would ever exchange with him.

"Who owns the Rimba?" I yelled to Riska, over the sound of the chugging engine.

"Oh, Mrs. Biruté. And Charlotte," she yelled back.

"Not Uil?"

Riska crawled across the deck so I could hear her clearly. "Maybe. Maybe in his name. And Pak Bohap and also I think Pak Aju, the owner of the Blue Kecubung. He's very rich because he owns a plywood mill. But Mrs. Biruté and Mrs. Charlotte — it's their place."

Biruté worked hard to assemble a Great Apes Conference near Camp Leakey in 1991 — there are testimonials all over the lobby — and now tourists pay hefty rates for Rimba rooms. Was this a venture into ecotourism? If so, it had given Biruté a strange bedfellow in Pak Aju.

"We'll do a walk here," Riska said, as we pulled up to the dock of the second, and newest, station. Pondok Tanggui was built to house the toddlers, the almost-ready-to-get-on-with-being-grown-up wild orangutans. Twelve or thirteen were living there at the moment, but the only resident we could see was Astra, a nine-month-old infant found in the Astra plantation near Kumai, where his mother had been killed. "He was only two months old when they brought him in," Riska told us. "His mama brought him out of the forest that was destroyed when they planted the palm-oil trees. They like to eat the palm because they're hungry. The workers spotted them and killed the mother, then sent little Astra to the park. Every night he cried, and the rangers had to take him to their own beds."

But Astra was in a room by himself when we went into the rangers' cabin to sign the register. Small and pathetic, he was crouched in a wooden box, clutching a rag. There was diarrhea on the floor and his eyes were enormous. They looked at us. From the doorway we looked back.

"He'd love some attention," one of the rangers said to Riska. There are two boys at each station. Here, they shared a bedroom with a CB that jabbered incessantly. The main room was bare except for a table and a few T-shirts for sale. In the back were shelves with food and a kerosene stove. There were two empty rooms, empty except for the wooden box with Astra and his rag.

"Want to give him his milk?" One of the boys was warming a bottle. Riska took it from his hand. "He's sick," the ranger told her in Indonesian. "From some tourist kids. Watch out or he'll get sick on you."

Astra did look sick. If he were human, I'd say he looked bleary-eyed, and why not say the same for an orangutan? He sat in a little heap and shivered. Two tourist children had picked him up and played with him a week before. They had been sick. Now we were invited to pick him up, and when I did, a little reluctantly, his small, hot body brought back the memory of holding my own feverish infant — Kristin so sick that I was using a cloth to drip water into her mouth; the two of us alone, her father somewhere else, but nowhere to be found. This baby had a small, bald head, enormous eyes, and tiny, wiry, grasping hands. There is not much hair on a baby orangutan, and what there is stands up straight so that Astra looked shriveled and yet surprised, surprised to find himself in such a place, and no wonder. I suppose a baby orangutan doesn't need ruffles and toys, but the place was too dismal for any of them, the boys or the baby. Just going out to the little porch with its heavy wire windows was a relief, although I took my turn holding the bottle and cleaning up vomit and feces and feeling guilty because, after we left to take our walk, Astra was alone again in the dark with his little rag. I remembered what Riska had said about the "breakdown" of the boy we'd met in Biruté's driveway. Indonesians don't like to be alone. Even with extra rooms, these boys slept together.

"Seen Dr. Muin?"

"Not yet, but he's coming."

That first walk, like all the ones to come, was through second-growth forest that has grown back on cultivated fields, or *ladangs*. Once, when there were more orangutans in Borneo than there were *Homo sapiens*, the great dipterocarps reigned. They offered their bark to be eaten and their branches for nests. Dipterocarps (named for their two-winged seeds) are still the overall giants of the canopy. The ones we walked beneath are young, but even so, finding a pathway between them is difficult. They drop their seeds onto soil so thin that, like medieval cathedrals, they have evolved enormous buttresses to support their fantastic height. What we saw, as we walked, were the shoulder-high buttresses, sinewy as old arms and covered with lichen and moss.

The orangutan's rainforest is profoundly slow in its rate of growth. An ironwood adds a yard to its waist every two hundred years, and some dipterocarps drop their first seeds at sixty. The erratic cycles of flowering and fruiting are the reason orangutans need a vast terrain. The fruits they cherish are figs, durians, langsat, and jackfruit, all of which have an odd adaptation in common: they appear not at the ends of branches but on the trunk or on low, thick stems, an adaptation that puts them in easy reach of orangutans who, because of their weight, are at a disadvantage higher up in the trees on the thinner branches. By eating the fruit, of course, orangutans also disperse the seeds. So the tree needs the animal as much as the animal needs the tree.

Indonesia's rainforests are so rich in species that only a small percentage of the plants and animals have even been named, and their interactions and interdependencies are still, for the most part, beyond our comprehension. The forest animals are utterly dependent on this habitat, of course, and most cannot survive in secondary forest with its much less complex ecosystem. Now, our bodies, too, began to adjust in small ways — though adjustment is a long way from adaptation. Different walking, different concentration, different focusing. We took in minutiae instead of whole. We gave our attention to place instead of time.

The trees are mainly trunks and leaves. Crowded together and throwing their energy into height, they are damp, living columns — graceful columns engulfed by twisting creepers and vines. The life around them takes place on various levels, just as it does in the sea. At the top, there is the canopy, where giant ferns and epiphytes use the trees to support their search for light, and where most of the animals live, relishing the freedom of movement and the abundant supply of leaves and buds and fruit and seeds. Civets, squirrels, lizards, and snakes — even some frogs — have adapted to an arboreal life and almost never come down. There, too, are the gibbons, leaf monkeys, and sometimes, female orangutans, who, at ninety pounds, are half the size of the adult males who forage lower down. Below, but still between earth and sky, saplings wait for a chance at the sun, and a few less-demanding species thrive. Where only the dimmest light penetrates to the forest floor, there's little undergrowth, only fungi and parasites, fallen leaves, fallen fruit, rotting wood, rotting corpses, excrement.

What's amazing is that a high percentage of both animals and plants in the rainforest are native to just one area. They live nowhere else on earth! Sixteen percent of all bird species live in Indonesia, and a quarter of these live nowhere else. Like the other animals, when they flutter through the dim light of the forest, they rely on loud calls to locate each other and elaborate camouflage as disguise. Constant rain, constant heat, constant growth. Fossilized plants prove that these forests looked much the same sixty million years ago. Long before that, as mountain ranges formed, as rivers emerged, as continents moved and heated and cooled, more and more species were isolated here, and with isolation they began to adjust, adapt, and specialize. The closely related monkeys, gibbons, and orangutans, for example, all live in the same treetops, but do not occupy the same niche. Alfred Russel Wallace, an early traveler to Indonesia and co-discoverer of the theory of natural selection, wrote, "If the traveller notices a particular species and wishes to find more like it, he may often turn his eyes in vain in every direction. Trees of varied forms, dimensions and colours are around him, but he rarely sees any one of them repeated. Time after time he goes towards a tree which looks like the one he seeks, but a closer examination proves it to be distinct. He may at length, perhaps, meet with a second specimen half a mile off or may fail altogether, till on another occasion he stumbles on one by accident." Two and one-half acres (one hectare) of Kalimantan forest may have 150 species, but only one example of each!

Unfortunately, this was not our experience, since we were walking in second-growth forest, and anyway we wouldn't have noticed because Riska's descriptions were more ecstatic than scientific, in spite of the fact that she knows an astonishing number of trees, their usefulness to local humans, and their Dayak names. Clearly in her element, she skirted the gluey areas where we got stuck as if she were made of something more ethereal than flesh. Where the ground wasn't soggy, it was slippery. Where it wasn't water, it was mud. But Riska was buoyant. Covered in long pants, a cotton hat, and a tough cotton shirt, she climbed over huge fallen trees and crowed with pleasure. "I was always a tomboy," she said with a laugh, pulling a bottle of water out of her red backpack and offering it around. "Climbing trees, getting into trouble. I was always into everything."

I admired her intensely. Exactly Kristin's age, she had many of the same interests, but their lives were so different! We were the ones who had traveled to get to her culture, yet the distance she must have traveled to meet us was at least as great.

During that first hike, I stepped on a nest of fire ants. I was wearing sandals. We were on our way back to the boat, and while I screamed and hopped and swore, there was nothing to do but wait for the stinging to go away and curse myself. Esta and Kristin were wearing proper shoes, but I was too hot for all that and insisted that my Tevas were sturdy enough for logs and mud. Insisted I was sturdy, too, but fire ants have the name they deserve, though fire had a whole new meaning after that. Esta and Kristin grabbed my hands and helped me hobble down the path, while Riska ran ahead to start our lunch. "In Hawaii, you'd pee on your feet!" Esta cheerfully pointed out, as I tried sand and spit. While Riska fixed mushrooms and carrots and greens with rice, Kristin did our laundry on the dock, and I soaked my burning feet with the underwear and kicked myself underwater, where no one else could see. The daughters were doing fine. They're fit and athletic. But I was becoming a drag!

The *Garuda III*, largest of all the kelotoks, pulled up a little past two o'clock, and Riska began to gnash her teeth. "It's against the rules to vary the feeding times," she said, "but Charlotte will ask them to speed it up so her group doesn't have to wait."

In fact, the rangers did begin calling the orangutans down from the trees when Charlotte appeared. For these twice-daily feedings, they prepare powdered milk and bring along a huge bunch of bananas. No rice. So that's the "new way," or part of it. "Only to feed them with their natural food to introduce them back to their original home," Riska explained. Wandering through the forest, the rangers were still loudly calling out names — "Monti!" "Doyok!" "Darmono!" — as, excitedly, we broke into groups of two or three people, dashing up paths, looking up into trees. Hoping to be the first to see the hungry arrivals, we kept meeting each other in the darkness behind roots and leaves. "Hello. Seen one yet?" "No." "Me neither." It seemed to me that Charlotte was annoyed at this sudden camaraderie. Perhaps it had nothing to do with Dr. Carey. Perhaps we, with our own kelotok and guide, were jumping into the fray too quickly and making it look too easy. The high-priced OFI tours are sold on the basis of orangutan inaccessibility and the provision of a guide. Yet here we were with exactly the same access to exactly the same facilities and without a $2,000 price tag. "Did you just come on your own?" some of the OFIers finally asked. "How did you get a boat?" "Are you allowed to sleep on it?" "Isn't that against the rules?" "So how much does it cost?"

Eventually three little orangutans came down from the trees, and this time, because we were in the forest, the sight of them swinging slowly from branch to branch was perfect, as if trees had been put on earth to provide aerial pathways. Here was the forest. Here were its inhabitants. "What if you lived in Eden?" a friend of mine once asked a roomful of students in a college classroom. "What would you do with yourselves?" We all thought a lot about that. "Drink tequila sunrises!" some of the students shouted. "But what then?" Now we, who had come back to it, simply stood in Eden with our cameras slung around our necks.

Connie Russell, the graduate student I'd met in Toronto who had studied ecotourism at Tanjung Puting, had concluded that most visitors hope to find nature in its "pristine" state and see the orangutans as photographic "collectibles." Tourists who view the orangutans as needy infants want photographs of themselves holding the fuzzy babies, and those more concerned with nature as pristine make sure to keep all signs of human surroundings out of their pictures. I found myself doing this, too, quite unconsciously, trying for pictures of the orangutan mothers and infants drinking their milk without any human legs or feet visible in the background. In fact, we were all bumping into each other, angling our lenses up and down and sideways, trying to pretend that no one else was there.

Milk was put down in a large plastic bowl where the mothers and babies began to gather. Taking us much more for granted than we could take them, they stood at the bowl's rim, most of them dipping their heads in, but a few of them, as the bowl emptied, picking it up in their hands. Taking turns. Primates differ in their ability to cooperate. Baboons, on one end of the scale, are severely hierarchical. Chimpanzees, at the other end, share their most valuable foodstuffs and keep only the ordinary, readily available foods for themselves. Most fascinating are the bonobos, once known as pygmy chimpanzees, who are dominated by the strongest females and share everything, including mates. With bonobos, *pax materna* reigns. Social scientists tell us that we became cooperative by learning to hunt larger animals in groups. The theory is that meat eating led to social life and even to speech, because hunting in groups required planning and fairly sophisticated communication. But bonobos, the great cooperators, eat fruit and leaves and twigs. No meat.

Hominids are forest-dwellers by design, according to Herman Rijksen, who began studying orangutans in the 1970s. But the design appears to have an odd adaptation, for "it has consciously sought to domesticate 'wildness' almost from the very beginning."

What of orangutans, who share much of our hominid design? The orangutan's stare is a sort of gaze, a meditation, a sense of time that is based on bringing long distance in close. An orangutan stares into a thousand shades of green and pulls one piece of it into focus. Is it ready? She moves her eyes a few inches and repeats the gaze. Closer. Closer. She cocks her head and begins again. Closer. Willing the fruit to readiness as a cat by the stove wills the mouse who lives behind it to come out.

Orangutans live in this state of meditation,

watching the object of desire, approaching hand over hand, when the *moment* is ripe. The moment itself is silently teaching the infant who rides on the mother's hip, and the other, the youngster who bounces ahead or follows behind. But it's the mother who teaches them how to touch, how to sniff, how to break through the skin. It's the mother who hands down this "culture." I use that word because learning is cultural if it occurs in one social group but not in others, if it follows a path of *relationships*. Among primates, the usual model for a learned behavior is the mother, which is why Biruté adopts the orphans. Not just to care for them but to teach them.

To be orangutans.

That night we attached ourselves to a floating rangers' station abandoned at the turn of the river. We'd moved north and then east, leaving the oily and mercurial waters of the longer course to take the clean branch that leads to Camp Leakey. The floating hut (known by the locals as Crocodile Homestay) has a sinking dock that provides the only place to stretch. The water here is unswimmable, as may be guessed by the name of our "lodge." In fact, nobody lived along this river until forty or fifty years ago because it is chock-full of crocodiles, or more properly, ghavials and false ghavials. And as far as we were concerned, the land beyond the dock was completely impenetrable as far as the eye could see. Thick forest comes right down to the river's edge. The dock — half underwater — circles three sides of the hut, and I circled it, too, climbing out of the boat and walking around it, looking for a private place to relieve myself — a search for the one thing left to desire? — but privacy was out of the question. The dock was too exposed, the forest too dense. Anang, who had managed to be invisible all the way upriver, appeared inconveniently at the prow of the boat. Riska chose my hidden corner as a perfect place to peel the vegetables.

There was no going ashore. The little dock and the boat were all we had. If we lived in the garden of Eden, I thought, it wouldn't be like this. We'd learn which branches are dependable and we'd glide right through them. We'd sleep in the trees at night in well-constructed nests. We'd investigate the animals and insects, fitting ourselves to their moves and purposes. We'd sample the fruits and notice the light as well as the texture and smell when they were good to eat. We'd teach our youngsters to notice these things and to find nuts and crack into them and to recognize foods that are dangerous. In this way, we would perpetuate our culture. We'd drop seeds after we had eaten, and these would be the seeds that must be exposed in order to germinate. We'd learn the plants that create comfort and those that alleviate pain. We'd live in the garden as part of it and teach our children how to survive.

As do the orangutans.

A Third Station

When we say, Thou shalt not kill, we do not understand this of the plants, since they have no sensation, nor of the irrational animals that fly, swim, walk or creep, since they are disassociated from us by want of reason.

— St. Augustine

"On your deathbed, would you rather be loved or respected?" We were tied to the solemn, sinking hut on our first night under the stars and under the blue wings of the *Garuda* when Esta asked this question.

Without hesitation, she and Riska voted for love, while Kristin and I agreed but then changed our minds. "No, respected," we both said. "People in the family will have to love me anyway," I added, "but from the rest of the world, respect would count for more."

Kristin narrowed her eyes. "We won't have to love you," she announced. A little warning.

We were an hour away from Camp Leakey, our destination and a site of pilgrimage for hundreds of tourists and scientists over the past thirty years. Now I began to wonder which Biruté would prefer, respect or love. Or maybe it's all just a question of degree, I thought. Maybe it's a question of to what lengths we are prepared to go to achieve one or the other.

Since it was our first night together on the boat, Esta and Kristin and I watched Riska's signals. While she assembled food and began to wash and peel and cut, we stretched, brushed our teeth, and took turns in the cubicle, pouring cold, black water over our bodies, as she had done to hers. Learning is specific to certain groups, not to an entire species. Riska was teaching the culture of the river, the culture of *Homo sapiens* living on small boats in the middle of the Sekonyer. She was calm, so we picked up that cue, and after dinner we females laid out our mats, and it was suddenly quiet. The *Garuda* was floating in the dark like a small, dead star surrounded by millions of trees. Yadi lit mosquito coils. River moved steadily under us. A wave of maternity swept in and there was no way to turn it back. In the darkness, stretched out around me, were the sleeping forms of five people half my age, and I was suddenly full of tenderness, as if I had hatched all these chicks myself.

The thing about being older is that the heart opens out. I have two stepchildren. Like Yadi, one of them was expecting his own child. Soon I would have a grandchild, which proves that at a certain point, if one is lucky, a child becomes yours for reasons that are not strictly biological. The blending of two families in the course of my move to Canada had taught me that, at least. Of course there was a time, not very long ago, when I would have viewed Yadi and Anang quite differently. They would have frightened or thrilled me, turned me on or off. My feelings then would have had to do with backs, arms, legs . . . although nurturing, too, is partly sexual. Nurture depends, for its strength, on vulnerability. It happens when we look at things with more compassion and less fear. I remember a wounded raccoon on a late-night Toronto street. Passing her in the car, I knew, in the dark, out of the side of one eye, that she was hurt, and I turned the car around and went back, sick at the sight and thought of a creature dragging herself along in spite of a broken back. She was mine, though had she been sound of body I'd have driven right past.

As the boys slept, and as I adopted them, I was not immune to their physical claims on me. But I was vulnerable, too. During the day, as I climbed off the bow to a dock, Yadi reached out to steady me. His care was evident in everything he did, from wiping down the deck to slowing the boat to point out a proboscis monkey high in a tree. So, too, with my girls. Gradually, I am changing places with my children. Gradually, I become the one who needs help, while they become stronger every day. Lying on my mat, I thought that Biruté could explain all this. Love must be the driving force, I decided, thinking of her nurturing fifty or more orphan orangutans.

But vulnerability of the self or the other comes into it.

Since learning is usually passed from mother to offspring, the duties of an adoptive mother to a child of another species must be daunting, indeed. Poor Biruté! When she accepted the uninvited task of raising orphan orangutans soon after coming to Tanjung Puting in the early seventies, she was trying to do something many experts believe is impossible.

Primates do not simply learn to copy an action. If that were the case, the action of mother and infant would be exactly the same. But that isn't how it happens. The details of the complex actions gorillas engage in to successfully get at stinging nettles are quite specific to each individual. What the gorilla learns is the overall form of the action, and this is true of *Homo sapiens* and the other great apes as well. It's possible for us to copy a plan because we've developed the ability to understand cause and effect. But Biruté, as one of the first surrogate mothers to orangutans, had to teach orangutans to do things she herself could not do. And this lesson was one she, herself, had to learn by experience. There was only one precedent.

Ten years before Biruté and Rod arrived in Borneo, Tom and Barbara Harrisson were living in Sarawak, where he was curator of a museum. In the course of his travels, someone handed him an orphan orangutan. He took it home to his wife and gave it to her on Christmas Day, putting it in her arms as she lay in bed. Later there was another orphan. And another. The mothers had been killed in lumber camps or on plantations, just like Astra's mother years later. How could the Harrissons refuse to take them in?

In her memoir, *Orang-Utan*, Barbara Harrisson describes their quandary:

> The babies that had come to us were in their helpless stage, when they are still entirely dependent on their mothers. Being adaptable and intelligent animals, they had accepted the human being as suitable substitute mother and started to learn and live in a human way. Was it humanly possible for humans to teach them the ape way — so that they might be fit to return to tree and jungle life? Contemplating my miserable anxieties and fears under the small shelter in the rain and darkness, I could not help being doubtful. Though being human, it is fairly easy to understand the ape way, to feel them being closely related as beings, as minds even, we live in fact very far away from their way. We have organized our lives and dependencies in such a way as to become useless in their jungle world. I felt that unless I learned to live like an ape first, it would be impossible to teach an ape baby.

Barbara Harrisson began learning everything

she could about the elusive primates. Since there was almost nothing available in the way of written information or research, and since he looked so much like a human baby, she reared the first orphan as if he were human, hanging a basket bed for him in the house and finding a young caretaker to carry him during outings. For this and later ex-captives, Barbara Harrisson came to believe that the forest was not an option. It was a belief that Biruté would try to disprove: that an orangutan without an orangutan mother/teacher will never survive in the wild. But what orangutans depend on is culture. Learned behavior. And a human can't teach an orangutan orangutan skills because we aren't part of that *culture*. Instinct may tell an orangutan to grasp, to hang on, to suckle, and to cry, but instinct doesn't extend to the subtleties of food supply or predators or weight distribution while swinging through trees. An orangutan not skilled in such things will starve or fall to its death or become dinner for a wild pig.

By 1970 Indonesia had begun its transmigration policy in which hundreds of thousands of Javanese have been sent to other islands with promises of cheap, cleared land or similar incentives. Most of the land they get belongs to the forest, and once cleared it is so undernourished that it's useless for growing anything at all, let alone food for a group of *Homo sapiens*. Still, this hasn't stopped the government. Or the settlers. Kalimantan has received nearly half a million of them. Much of the ecological damage in Borneo is done by desperate Javanese working for lumber, palm, and mining companies. Squatters cut trees, sink pits for gold, work in mines and on commercial plantations. These activities are savaging the environment, and the result is more and more orphan orangutans.

I pictured again the first arrival of Biruté — she and Rod motoring up the river in 1971 with all of their lives before them, bringing a few magazines, a few clothes, and more consequence to the forest in their sampan than anyone would have imagined it could hold. About the time Biruté arrived at Tanjung Puting, the Indonesian government began to enforce a standing prohibition against owning orangutans. Because of transmigration and lumber concessions, the forests were being cut down so rapidly that two things were happening: overcrowding of the orangutans due to the destruction of their environment and confiscation of the orphans who fell into willing or unwilling hands. After confiscating several, the forestry department in Kalimantan asked Biruté and Rod to take them in. Could they be rehabilitated? So Camp Leakey — twenty square miles of trees and trails with a partially cleared patch for cabins — began adopting orangutans. It was a crucial, perhaps irrevocable, departure from her original mission, which was research into the lives of wild orangutans. It led to the adoption of more than a hundred orphans over the next twenty years, to the founding of an international foundation to support her work, to the politicizing of Camp Leakey and Tanjung Puting, and to the flourishing business of ecotourism, bringing money to the Sekonyer River.

These were my thoughts as I tried to sleep. But my feet were itching. Was it the fire ants? I was awake off and on all night, and as I lay under the thin blanket, rubbing my feet together like a giant insect, I thought about the third station, Camp Leakey, and why it's a destination for all kinds of reasons, not least of which is that it's the end of the road. Beyond it, the river is impassable, at least in a kelotok. One does not pass by on the way to someplace else. Now, I was going to see this legendary place. From the ground up. And when I set foot on dry land, I would be wearing shoes and socks.

The next morning, as we pushed our way through the thick reeds that choke the branch of the river that is the only path to Camp Leakey, there were no other boats; we could see nothing but green and sky. They clung to us, those elements, above and around. Occasionally, the boat had to come to a standstill and reverse, for we would have hit a bar of sand covered in plants. "Lots of crocodiles in this part," Riska said once. "I remember one time . . ." but no one was listening; we were straining our eyes. An hour or so from the old floating hut where we had docked, the tip of a wooden fire tower appeared, and I knew one of those rare moments that make traveling in a straight line feel like traveling in a circle. I'd seen this landmark on a slide while I sat in the classroom in Long Beach, California, trying to imagine myself doing this. And here I was, complete with two daughters and a Dayak guide.

We climbed out of the boat and stood on the dock, which is so long that it vanishes at one end, the way a sail does on the curve of the earth. Once this dock was crossed, though, Camp Leakey looked simply clean and settled and swept. It had the look of a deserted summer camp.

The long dock turned into several well-beaten paths leading to painted frame cabins. In one of them, there was a ranger to take our permits and a book for us to sign. In the silence, we stood in the hot shade of the trees looking up. There was no sign of life in the branches. On the ground, no sign.

Camp Leakey was named for the man who sent Biruté to Tanjung Puting a few years after she met him at UCLA. She was a student. He was a visiting lecturer and, after thirty years of digging in the Olduvai Gorge, the most famous scientist in the world. "I want to study the orangutan," she told him. Both of them were diligent spinners, both of them were casting webs. He asked her to visit him the next day, and said he was staying with friends.

The meeting is more or less legendary because

it describes the two spinners so well: the old roué was shuffling a deck of cards when she arrived. As they chatted, he laid the cards facedown and asked her to identify the clubs and spades. Biruté replied that she couldn't know which cards were black but she could see that half of them had bent corners. At that, Leakey smiled. He had been looking for a woman to send to Borneo. He had found his third angel.

He believed women were better observers than men, but it took a particular kind of woman to observe great apes in the wild. She believed men were less interesting than orangutans, but it took a particular kind of man to gain her access to the forest.

They had caught each other.

Perhaps they shook hands.

Tanjung Puting, where Biruté established her research station, was a nature reserve in those days, created to protect the diverse species of its swampy, lowland forest environment. Now, near the rangers' cabin, a sign reads:

Don't take anything except foto.
Don't leave anything except trace.
Don't bring anything except memory.

So, this time, I would take a "foto" of Biruté's house, where she lived until she moved to Pasir Panjang. "Which way is it?"

As Riska took the lead, she told me about a boy in Pangkalan Bun who once reported to

Biruté that two baby orangutans were being kept by a neighbor, a man who had acquired them as pets. No one in a village society wants to be an enemy to his neighbors. This sort of reporting took lots of training and reward, although by law only the PHPA or police could confiscate orangutans.

On this occasion, Biruté sent her personal assistants to the man's house, and they demanded that the orangutans be handed over on the grounds that they had been obtained illegally. "That orangutan owner tried to have my friend killed," Riska told us. "Yes. The boy was threatened and had to leave town. I took him to Dr. Biruté to ask for her help, but she said she couldn't help with the problem; it wasn't her responsibility. As far as the man with the pet orangutans goes, he said he would have given up the babies if they had asked, but he was no thief and did not like to be insulted."

When the government decided to put some teeth into the law making it illegal to buy, sell, or own an orangutan, somebody had to take care of all the pets who were confiscated. It was a thankless job, but since Biruté and Rod were already managing fairly well with the orphans taken from lumber concessions and palm plantations, Tanjung Puting seemed like the logical place for future ex-captives.

The story Riska had told us of the boy who reported his neighbor is only one of a thousand stories that have Biruté charging into battles where lesser angels might fear to charge. Often her efforts got her into trouble with the locals. Sometimes they got her into trouble with the PHPA. She was getting the job done, yes, but in her own way. Meanwhile, more and more tourists came to see the orangutans. There were magazine articles, television documentaries. Then the orangutan map expanded. Things got political.

When six baby orangutans were discovered at the Bangkok airport tightly packed in sealed crates labeled BIRDS, the date was February 20, 1990, and the case made international headlines. They'd been smuggled out of Borneo and flown in from Singapore, and they were found by Thai Airways officials who X-rayed the crates when they heard what they thought were the anguished cries of human babies. And no wonder the cries were anguished. Three of the young orangutans, having been packed upside down, were in desperate shape. Their lungs were full of fluids. None of them had been fed. All of them were sick. But since the only law broken in Thailand was one that demands proper labeling of goods being shipped, if the crates had been labeled ORANGUTANS, the animals, although officially considered to be endangered, would have been put on the next plane without any questions. Two or three or all of them would have died on the way, but this is a normal statistic in the world of animal smuggling. This time, however, they were eventually sent to Biruté in

Tanjung Puting, and there are different versions told about what happened next.

Like drugs and gold, orangutans are commodities. They're worth thousands of dollars, and since they can't legally be taken from the wild, their value has only increased. To the Dayak, who call them *mias* or *mawas*, they were "wild men" guarding the forest and competing with less wild men for its fruits. Their heads ensured the fertility of the fields just as human heads did. (The Dayak still tell of red-haired men who come out of the forest to eat the ripe durian.) So they were hunted. In small numbers. Even now, orangutan heads can be found for sale to tourists in Samarinda, and they occupy favored spots in the rafters of many a longhouse. But now *live* orangutans are desired by zoos, by circuses, by Hollywood. They're kept by Indonesian generals and housewives in Taiwan. They're purchased by scientists for medical, pharmaceutical, or psychological research. In Asia, they became fashionable when Dutch colonials began keeping them as pets. They became fashionable because they are so much like us, and as a result, they are smuggled, traded, and sold, although the capture of a live orangutan usually means the murder of its mother, and the mortality rate during travel and captivity is so high that obtaining one live orangutan often means the death of eight or nine others. Orangutan females carry their young in their arms or on their bodies and this slows them down, so killing a mother is the *only* way to capture an infant, an infant whose only reality is mother, and who watches as she is decapitated and hacked into pieces. Then, orphaned and completely dependent, he is gathered into the arms of the murderer and taken away from the forest.

In 1993 a meeting of the Captive Breeding Specialist Group of the World Conservation Union was convened in Medan, North Sumatra. Knowing that a wild orangutan will not conceive before her thirteenth or fourteenth year, and with an average birth interval of eight years and a life span of about forty-five years, they estimated that a very small increase in mortality of five adults per one thousand would lead to extinction within three decades!

It's for this reason that poachers and animal smugglers are taken so seriously. By some governments, at least. Although there were other people involved, the International Primate Protection League and the U.S. government spent five years pursuing one man in order to bring him to trial for his involvement with the Bangkok Six and make an example of him. Eventually, the U.S. Fish and Wildlife Service accused him of smuggling the six orangutans out of Indonesia. After a battle of wits and mutual recrimination, he was sent to prison in 1995.

Meanwhile, at least three of the Bangkok Six had died in the national park of Tanjung Puting, under the supervision of Biruté. But a California woman named Dianne Taylor-Snow was unofficially blamed. Who killed Cock Robin? What happened to the other three orangutans? It depended on whom I asked.

These days, the most frequent murderers of orangutans are not poachers but planters, who shoot orangutan mothers caught in flagrante nibbling at palm-oil trees. Caught nibbling because they love the heart of the palm and because the forest that has been cleared for the plantation is what feeds and houses orangutans. Cut down the trees and orangutans can't eat. Shoot the mothers, and the species can't reproduce. And without a viable population of this arboreal primate, it's possible that the remaining forests of Borneo and Sumatra may not survive, since orangutans help keep the forest effectively diverse.

Camp Leakey has a little cemetery near the dining hall. There are carved wooden markers, a plot of mown grass. Out of the center rises a flagpole that seems to diminish the graves underneath. We wandered past it on the path to Biruté's old house. Here, she lived with Rod Brindamour and later with Pak Bohap. Here she kept records of what wild orangutans were eating, what ex-captives were doing. Births. Deaths. *I have systematically collected data on all aspects of orangutan behavior such as diet, foraging patterns, locomotion, and postures, day ranges . . .* The house is white. There is an unpainted porch. It is possible to stand on the porch and look in, but what meets the eye is a room where no one has lived for a very long time. *Our conservation work has involved rehabilitating ex-captive orangutans to life in the forest (and thus breaking the back of the orangutan trade in the province), patrolling the National Park (also destroying settlements and chasing loggers out), confiscating fishing lines and nets, training Indonesian University biology students and Forestry department personnel and educating the local public . . .*

"It's not what I thought it would be," said Esta, as I lifted my camera, pointing the lens through the wire grille that covers the window screens.

"Would you rather be loved or famous?" Kristin said.

"I thought it was loved or respected."

"I guess it was."

The empty room was in focus but, again, I felt ashamed of myself and put the camera back in its bag.

The Good Doctors

By the dock, climbing out of a water taxi, we saw a slim man in the khaki uniform of the PHPA. "It's Dr. Muin," Riska announced, and we padded down to greet him. This man who dealt with orangutans in extremis was young and slight and as animated as a boy. At twenty-five, he did not look the part of a doctor to primates twice his size.

He apologized for having missed us the day before. "I was doing water samples for Dr. Carey," he explained. "Now I want to get the results up to her, but I don't have a way to get there."

"I have a date with her at twelve," I said. "Why don't you come with us?"

Dr. Muin had come to check on Purwasih, one of three young females who cling to each other so fervently that I later dubbed them the Twisted Sisters, and there they were when we hiked back up the path with him. With some calling on the part of the ranger, they were rolling across the patch of ground in front of us like a furry, multilegged creature from Oz, never letting go of each other for a second.

But in order to check Purwasih, Dr. Muin had to separate them.

He first tried to distract Augustin, the oldest and most maternal of the three. Then he tried to entice Purwasih, the little one, away from her. When neither tactic worked, he tried Tata, the middle one, who was more subdued than the other two and finally allowed herself to be distanced from them by two or three feet. That was no good either. Within seconds they had all rushed over to an old wheelbarrow and climbed into it, as if into a fortress. Their alarm was painful to watch, but at the same time it was morbidly fascinating, for here were three young primates who had adopted each *other*. It gave me something new to think about.

Now one of the rangers produced a discarded packet of dry noodles from the trash — the kind of packet that contains a little envelope of soup mix and makes a good instant lunch. Forbidden under the "new order," noodles are adored by orangutans, and Augustin was momentarily seduced. As she reached out, Dr. Muin swooped in, grabbed Purwasih in his arms, and ran to the rangers' hut, hurriedly locking the door behind him.

Screaming, Augustin ran after him.

I thought about the way I was tricked into having my tonsils out by a promise of ice cream — how betrayed I had felt — and I was a child with a secure home life. (I was also a child who picked up wounded birds and peeled dead turtles off the street. I had both a hospital and a burying ground.) Little Augustin clung pitiably to the wire covering the window screen and cried even as she was eating the forbidden treat. These animals had been terribly traumatized by the deaths of their mothers, and on the other side of the wire, Dr. Muin was putting a needle into the one thing that made Augustin feel safe in the world. So there we were on the porch of the cabin with a distraught orangutan and my tendency to save things. When Augustin had finished the noodles, I reached over to take the paper packet away. It was empty; I didn't want her to swallow it. I wanted to help. But she sprang off the screen and came after me. "Mom! Quick!" Kristin shouted. It was not the last time she would use those words on this trip, but it was the only time there was real emergency in her voice. "She's going to bite you!"

Indeed, Augustin was looking extremely toothy as I took four or five terrified steps backward, clutched my camera, and dived off the porch, just the way they do in old westerns, landing in the dust flat on my back. The pitiable orphan had become a raging ape. The transition had taken only seconds. It took me much longer to get my heart quieted down and get up off the ground, rubbing my sore backside.

In the locked cabin, Dr. Muin had given Purwasih an injection of an antibiotic. When he opened the door and released her, she jumped straight into Augustin's arms and they rolled down the steps to find their adopted sister, who sat shivering on the sidelines near a dusty tree.

Dr. Muin descended, too, rolled down his sleeves, and announced that we could now be on our way upriver.

Once he settled himself on the lower deck, leaning his back comfortably against the sloping sides, he seemed happy to talk, and we both managed to shout above the motor's racket. Riska began to fix lunch, and Esta and Kristin went above. The galley was too small for more than one person at a time, and Riska was the only one of us who knew her way around the stove and equipment, although sometimes we helped her slice things up in order to ease our pangs of conscience. It was strange having someone else — another woman, especially — feed us. I think I adapted more easily than Esta and Kristin, who were after all Riska's contemporaries and uncomfortable with any symptoms of elitism we might be showing. Relax, I kept telling them. How would you like it if somebody came into your office and started arranging things on your desk? This is Riska's profession. Still, we felt self-conscious and unaccountably useless.

"So, you came to see Mrs. Biruté," the doctor said.

I told him I knew that she was in Canada, but I was trying to measure her effect on the lives of the animals she was trying to save. "And the river. And the trees. What happens when we . . . I'm interested in saving," I said. "I mean, in why people do it and whether her work at Camp

Leakey is making any difference to the species or only saving a few individuals."

It was a new way to phrase my interest, but Dr. Muin was shaking his head. "But she has no permit to work here. Not anymore."

"At Camp Leakey . . ."

"No. Not there or anywhere else."

I had heard hints of trouble, but no permit? "What's the permit for, exactly?"

"Well" — he took a sip of the water Riska had handed out to him — "there are two. One for research. One for rehabilitation."

"Which one is revoked?"

"Both of them."

I let this sink in.

Right away, he had much to say about "Mrs. Biruté," including a bitter account of a guide named Fuadi, who is hired by journalists to arrange for access to her. He said Fuadi and Mrs. Biruté work together — she charges the journalists a lot of money for each interview and gives him a slice of it, refusing to work with anyone else. He said, "Whoever heard of a scientist charging money for an interview?"

"But why exactly did she lose her permits?" I asked, wondering if the complaints against her were any more substantial than this contempt for her professional manners. "And how long ago?" In Long Beach, a paper with the following information had been handed out to the students:

Barriers to Galdikas' presence in Tanjung

Puting National Park include: 1) poachers who illegally take forest products, 2) timber barons who would like to see her removed from the Park in order to obtain the virgin timber, 3) gold miners who use destructive methods to extract the precious metal within the boundaries of the National Park, and 4) some other influential individuals who are jealous of her long-term commitment and good relations with the local community around the National Park.

Dr. Muin's eyes are bright and they seem not to blink. "For one thing, the park requires reports every three months and she never sent them. We had no idea what condition the orangutans at Camp Leakey were in or what research she was conducting, if any. Camp Leakey is not her property; it's a small part of the whole park. Sometimes we suspect she is not doing much except impeding our progress at rehabilitation and making money from tourism. She refuses to turn over the orangutans people bring to her house, although legally they belong in the park. She keeps them there at her own house, which is not only illegal but sends an ambiguous message to other people about the laws she wanted us to establish in the first place. So many scientists are against her." Then he said, looking down, "She lives in a world of her own."

I asked him what, exactly, in the context of Tanjung Puting and the great rainforest around us, that could possibly mean.

"For example, the first time I discussed rehabilitation with her we were at her house and she said, 'The most important factor for orangutans is food.' Really! She doesn't care about disease, although Willie Smits at Wanariset stresses this factor. She denies it, which is dangerous for the orangutans. Do you know about Wanariset in East Kalimantan? That is where real rehabilitation is taking place! Food supply is simply not more important than disease or weather or human beings. Also the people who work with her are not scientists. They should be trained in science. In this park the rangers are not allowed to give noodles or rice to the orangutans any longer, but her workers still do this. The policy has to come from the top. You have to have basic science if you really want to work for orangutans."

"Her workers?"

"Well, yes. There is still some of her staff at Camp Leakey."

"That seems odd, if she doesn't have a permit to work there. When did all this happen?"

The young doctor glanced away.

"Oh, I don't know — 1993?"

I saw him as a newcomer, as one who did not have the whole picture, as one who was full of opinions, the way young people are when they come from the city to a place like this. I wondered how much of his diatribe had originated in Jakarta. It's taken a long time, but today the government of Indonesia officially acknowledges the value of its rainforest, especially, as they put it in their reports, "with regard to the genetic diversity of major timber tree species and other useful organisms." In fact, very few areas of lowland and swamp forest have protected status, and virtually all accessible lowland rainforest has been irrevocably damaged. The damage is caused by exploitation of timber. And the ecological structure of these forests, once exploited, is forever lost because with primary forest, duplicate reforestation is impossible.

It's the loggers who make it easy to finish the job of destruction they've started. By building roads, they open the forest to plantation development, settlement, and even poachers. They make things easy for *Homo sapiens* by opening up its most remote and secret recesses. But they make things hard for orangutans and other members of the forest. As their habitats are degraded, orangutans begin to roam into human ones — the very palm plantations and gardens that have been thrown up to make money from land that is otherwise worthless. And since law enforcement for wild animal protection is extremely lax, any confrontation between human and orangutan is likely to lead to an orangutan death. Baby orangutans fetch anywhere from $100 in Jakarta to as much as $25,000 in the United States. If the dead orangutan is a female, the sale of her baby or even her skull can change the fortunes of a family.

I asked Dr. Muin what he thought of Biruté's

idea of establishing a new rehabilitation center in Hawaii.

"What center is that?" Dr. Muin was visibly shocked. "Home for orangutans is Indonesia!" he exclaimed, nearly knocking over the glass of water. "If she really wants to rehabilitate the orangutan, she has to think about how to repatriate it and conserve it here, not try to make a new area in another country."

Since we were having to scream over the roar of the *Garuda*, we resorted to the lunch Riska now served and continued the trip in silence.

I was two hours late for my appointment with Carey Yeager, but I was sure she wouldn't mind. After all, we were on a river in the middle of nowhere. We were transporting Dr. Muin and the results of his water testing. And what is time but a river itself? Although she isn't mentioned in any of the OFI data, Carey Yeager worked with Biruté for two years, between 1980 and 1982, managing Camp Leakey in Biruté's absence on two or three occasions for several months at a time. A primatologist, she knows the staff — the ones who have disappeared and stayed — and she knows the orangutans — the ones who have disappeared and stayed.

Our first sight of her had been impressive. Standing on her own long dock, surrounded by the Trekforce team that had arrived with us on the little plane, she looked clean and pressed, a woman who was definitely in charge of her surroundings. The Trekkers had paintbrushes in their hands. "Sorry we're late," I'd called up.

"Yes, I don't have much time. I have a group coming in half an hour." So apparently tardiness was not acceptable. I saw Carey check her watch. But she'd climbed aboard the *Garuda*, accepted a cup of tea from Riska, and now she seemed ready to answer questions. I asked her about the success of Biruté's efforts to save orangutans. "A lot of them are gone," she said quietly.

What did that mean?

"Gone," she repeated. "Who knows where?"

Carey worked with orangutans and then established her own camp two hours upriver from Camp Leakey. Now she studies the only protected population of proboscis monkeys in the world. With a furiously busy life, half at Fordham University and half on the Sekonyer, she'd taken the time to respond to my letters, to reserve the *Garuda*, and to send useful advice: "Do live on your kelotok. It's the best way to travel here. Please don't take a speedboat. They kill turtles and even monkeys." Now we sat with our legs stretched out in front of us on the blue floor of the boat she had obtained for us. It was two o'clock and hot as the pan in the galley in which our *nasi goreng* had been fried.

"This group," Carey said, as if to explain her tight schedule. "I wasn't expecting them, but now they're here. Why don't you come up at three? If you want to hear my little talk to the group, you're welcome to come up now." She looked at Esta and Kristin. "All of you," she said.

I felt I had made a friend. I told her we'd like to do that, and she crawled through the forward door and then stood up and jumped over to her well-built dock. Trekforce had extended it several years before, and now they were spread out along it, painting huge pieces of plywood a bilious orange. There were streaks of it on their clothes and shoes and hair.

This fourth station, officially called Natai Lengkuas, must resemble the early Camp Leakey. Both have long wooden docks bolted to the river bottom and leading across the intervening swamp to higher ground. The docks are surmounted by the usual national park logos and purely symbolic gates that designate the beginning of something worth visiting and set the world of nature apart, dividing it from the commercial world of the river. The docks may be deserted or surrounded by kelotoks, water taxis, and sampans. Their surfaces may be starkly empty or covered by people washing clothes or resting or bathing. Men in this part of the world strip down to underpants — usually briefs in pleasant, surprising colors. Women wash themselves furtively, exposing only face and arms and lower legs. Swimming is possible only off the dock at Camp Leakey, where there is, supposedly, little chance of meeting a crocodile. But a station is a station in these parts, and from the water they look pretty much the same.

The group that had turned up to hear Carey's talk was a small one led by Scott Atkinson, who worked for Bolder Adventures when he wasn't traveling on his own or working on his master's thesis on ecotourism. This particular group was not part of the OFI tour, but quite separate. "For my thesis I wanted to write about orangutans," Scott told me later. "Especially the ones who rape. No, really, it happens. Mostly with the adolescents. It's a big subject, but I couldn't find enough material. There isn't much research. With all her talk about being a scientist, Birutė isn't doing anything."

Scott is very dashing, and we all perked up when we saw him standing under the trees listening to Carey's lecture. I could feel Riska smile, although I was behind her. She went across to say hello while Carey explained to his small throng that proboscis monkeys are like arboreal cows, with sacculated stomachs that break down leaf fiber and toxins so they can eat things other primates can't eat. Ripe fruit gives them bloat and they can "blow up," she told everyone gaily, but the seeds of unripe fruit are a proboscis delicacy. This is not bad for the forest, she said. Proboscis monkeys aren't seed dispersers like orangutans, but they assist biodiversity through seed predation because the seeds they eat are from the most abundant species. So their predation acts as a reproductive constraint on these species, allowing other ones a chance to reproduce.

It never fails to lift my spirits, the great and complicated pattern that is exactly as it should be, and I decided I liked the proboscis monkeys for their usefulness, although as a large-nosed person I had imagined that I would love them a little for their noses, which are large and bulbous and which, in the males, hang down very foolishly. Then Carey moved into higher gear. She was explaining that the proboscis is a threatened species existing only in riverine and coastal areas of Borneo and that the threat they face is habitat destruction. Just like the orangutans. "The *lumber* industry is the problem," she said angrily. "This park has the only viable population of proboscis monkeys on the island that is protected," she said, pointing out that Kutai Park, by contrast, is badly degraded by fire and logging.

This led to discussion. What about the river? Isn't mercury poisoning it? Isn't that the first priority? What about the money tourists bring in?

"Your money doesn't reach the animals," Carey said flatly. "It doesn't even get as far as the national park, except for the tiny fee you paid to get a permit. You can't personally give the park or the rangers money because by law they aren't allowed to take it, which is probably a good thing. The average annual wage around here is U.S. eight hundred or a thousand dollars a year for a family. And there are somewhere between a hundred and fifty to three hundred families in Tanjung Harapan village alone."

The money tourists bring in probably equals

the money locals earn logging or mining, she said. "That's quite a lot of money. It just doesn't end up in the right hands. In the long term, trees and gold are being played out, but tourism can continue." Carey said she's working to find local products people could produce for income.

"I've even been talking to people at the Body Shop," she said. "We have so many herbs here, and we could grow even more. And reforestation. That could provide employment. We could plant local species plus commercial fruit trees. Last year there were twelve hundred and eighty visitors to the park, but with a few improvements, that could be tripled."

Tourism. Ecotourism. When Carey talked about boardwalks in the park where tourists could walk without disturbing the forest, I looked over at Scott and wondered what he was thinking. For the first time since I'd arrived in Borneo, I felt hopeful. Carey seems to infect people with her energy and determination, and I thought, What a pity the OFI group can't meet her. Things are possible, she seems to be saying. Just look; just listen; just watch me. As a guide working for Bolder Adventures, Scott can't bring OFI groups here, although he says he always brings other groups. It hadn't occurred to me to wonder about the dollars tourists spend for OFI tours. I'd assumed that a good part of the money charged by the OFI and other tour groups went to the park to help the forest and the orangutans. I'd assumed that a chunk of the money I had paid to become a member of the OFI did the same thing.

While Carey talked, her husband added exclamation points to everything she said. They've been together long enough that they finish each other's sentences, although sometimes a verbal detour takes place: "No, that's not it. Not what I meant, Trevor." When he lit a cigarette, she made a face and waved her hand around to clear the steamy rainforest air. He laughed and moved away. At home, he's a Unitarian minister. When they come to Kalimantan, he becomes manager of Natai Lengkuas.

Neither of them discussed their work in the park except in a general way. Trevor even insisted that the work they do can only be hindered by attention. "We're not in it for that; it gets in the way." Carey had agreed to meet me several weeks before, when I'd sent her the same letter by fax and through the mail. But when the public talk was finished, she became reticent. The aversion to public attention was serious. She told me a story in an effort to explain her feelings. A friend of her mother's had published an article in the hometown paper based on letters Carey had written to her mother. "And she said, get this, she said that I wash my clothes on a rock! Which is really ridiculous, I mean, come on! There's not even a rock close to here! She just wanted to make me exotic."

I laughed, and within moments she launched into the complexities of life in the forest among the orangutan rehabs. The dam had broken. But only temporarily.

That afternoon we went up to the office. It was cooling off just slightly and time for the "official" interview to begin, the one that I'd requested by mail. The office had a huge plywood table with benches around it, and we lined ourselves up on one side of it, Riska, Esta, Kristin, and I, as Carey and Trevor sat down on the other side. Scott came in and hovered at the edges, then sat on our side. "I was wondering . . ." I began, uncapping my pen.

Trevor said, "Not until you sign this," and put a document down in front of me. When he laid matching pieces of paper in front of Kristin, Esta, and Riska, I glanced at mine and began to sweat in earnest.

Kristin nudged me. "You can't sign this," she whispered.

"I can't sign . . ." I said.

"Then the conversation is finished."

This struck me as worse than unnecessary. It seemed downright unfriendly. Here we were in the rainforest of Borneo on a river whose name I could barely pronounce and I was being asked to sign a waiver stipulating that I would not publish *anything* about my experience without Carey and Trevor's permission! "She could shut us down in ten minutes," Trevor said. "That's the reason. We have to be careful not to get ourselves into print. She has friends in the highest places."

I said, "Birutė? But she doesn't even have a permit to work at Camp Leakey anymore," and tried to focus on the words on the paper in front of me, still hearing Kristin's "You can't sign this" and Trevor's "She could shut us down . . ."

Carey said, "I wrote to you about this," which was true, but in the middle of our friendly meeting, how was it possible that my hosts had become so untrusting?

We primates are hierarchical. There are ways of communicating who is dominant and who subordinate. One chimpanzee will crouch and make panting and grunting sounds while another pulls himself upright and makes his hair stand on end. One may swagger, another may kiss. An open mouth can be appealing or scary, depending on the tilt, but certain things are always threatening.

I noticed that Esta, Kristin, Scott, and Riska had left and wondered if they were being discreet or if they were outside judging me. "You're free to write anything that's true," Carey said. "And doesn't hurt our program. That's a promise. But some things are off the record."

I signed.

What we talked about, then, was park politics. "One third of the park budget goes to orangutan ex-captives," Carey said soberly. "Which is out of proportion compared with what gets spent on the wild population."

"You said a lot of the ex-captives are gone. What does that mean? How many orangutans are we talking about?"

Carey didn't hesitate. "I have some statistics. There aren't many records available for releases during the early years," she said, referring to the practice of taking rehabilitated orangutans so deep into the forest that they are unlikely to return. "But I know about the time between 1980 and 1982. There were twenty-six individuals at Camp Leakey during that time, eight males and eighteen females. Of those, we know two are dead, eleven are still around the camp, three were forcibly removed and taken to another part of the park because of attacks on humans. One of those is presumed dead. And ten disappeared, at least two of which are presumed dead. If they're taken into the forest, we never see them again. They may be killed by wild pigs or starve to death. Sometimes if they get close to camp, we can see that they are ill or have deep wounds from encounters with wild orangutans or predators. It's nice to say they've returned to the forest, but the truth about rehabs is that no infant taken from its mother and kept in a cage for some time is ever going to be anything but mentally damaged, no matter what happens. Put a human baby in a closet for a couple of years and then raise it and tell it to go be normal and take care of itself, go get a job. It's impossible. The damage is too great."

Thinking of Purwasih, I said, "Unless another orangutan adopts it."

"A wild orangutan, yes. But not a rehab. They

can't teach a baby how to forage and survive. They don't know themselves."

"Because they have to be taught all the specifics of eating?"

"Right. But tourists don't want to know this."

"It sounds like Biruté is doing more harm than good!"

Trevor had taken the waiver from my hands and deposited it in a filing cabinet. The air was cooling at last, but I still felt shaky, as if I had seen a wall thrown up against the current of the famous river. As if that current were about to rise. My two hosts were nodding their heads.

It was late in the day when I got back to the boat, where I found Esta and Kristin reading. "They invited you guys to go out with the Trekkers tomorrow morning if you want to," I said.

Kristin said, "Did you sign the waiver?"

Sinking to my knees, I made myself small, feeling that I had let her down. I had proved that other people can control my life, that I couldn't stand up for myself. Or maybe I had set Kristin the onerous task of inventing a family standard. Maybe that has been her role all the time, through break-up and divorce and everything that followed. She was looking at her knees, avoiding my eyes. "You won't be able to write *anything*."

"I want to go," Esta said. "It sounds great. With the Trekkers. What are they doing?"

"You'll have to sign the waiver."

"I don't care. What are they doing?"

"Measuring trees."

"Sounds fabulous."

Kristin said, "I'm going to the mining place, like we said."

We ate our beans and rice and vegetables in silence. After dinner, I went up the dock and the steep hill beyond it in the dark. Carey and Trevor had invited me, and this time I was shown to a small sitting room, where we shifted our legs, our eyes, our faces. Within our small triangle, Trevor and I, for different reasons, were all wariness and concern. It was incredibly hot, the air was still, and we sat in the small room and drank warm Cokes like reasonable people. There was a rattan two-seater and a rattan chair. A coffee table. For years both the office and their living quarters were in the small house they still inhabit nearby. Now there is this difference. The Trekkers have built something durable. They have worked for the good of this place. The skills that are needed here are often mechanical. Trevor himself has managed to fix up a solar panel, and when that doesn't work for the small computer they bring with them on the plane, he runs it off a car battery. Thinking that they must sometimes miss Camp Leakey (and perhaps that the men in these marriages get too little credit), I told him that I'd walked on the log paths Rod Brindamour built twenty years ago, that they were still holding firm, and he said appreciatively, "I'd trust anything he touched." He said it with such admiration and sincerity, I thought what a loss to Biruté Rod Brindamour's leaving must have been.

We began to talk about rehabs again. "We have third-generation welfare mothers in this park," Carey said. "The babies come to Camp Leakey and stay, and then their babies stay. Of course the infant mortality rate is very high because these orangutans don't know how to be mothers and they reproduce much more frequently than a wild orangutan would — sometimes every three or four years as opposed to seven or eight. This is completely unnatural and simply increases the dependent population, which puts further pressure on the wild ones."

"Has anyone considered birth control?"

At this Trevor and Carey spoke in unison. "Oh, sure!" Then Trevor continued, "But it's a very tricky question. If it isn't handled right, a lot of people would get upset. Biruté, for one. She probably wouldn't like it and she'd start a fuss."

"Like sterilizing welfare mothers?"

"Yes, but they could use Norplant, which works for five years or something like that. Exactly what a female needs, if she's going to make it in the wild, is a fighting chance to survive. She has to learn the ropes. Let her have the chance and then let her reproduce when she knows what she's doing. I think this could be understood. But there are two kinds of tourists — the fuzzy-monkey ones and the ones who prefer animals in their natural state. They play a big part in all this."

Birth control for an endangered species? Wanting to earn my stripes, I said, "Rehabs may seem like fuzzies, but they're more dangerous than they look." Then I related the morning incident with the Twisted Sisters, Purwasih, Augustin, and Tata. I said that I, for one, had no idea such a small animal could really do serious damage, that Kristin yelling "Mom, she's going to bite!" had probably saved me.

When Carey described her misadventure with an orangutan named Rico, I knew I'd been right to dive. Rico was an adolescent male at Camp Leakey who was "always a biter. He bit lots of people. Thirty or so." One Camp Leakey morning, as Carey told it, Rico stole a bar of soap from a tourist, and the tourist tried to take it back. When Rico attacked her, Carey, who was bathing nearby, grabbed a stick and began to brandish it. "With primates, everything is a test of power. Strength. That's what primates do. They are always busy establishing hierarchy."

Trying to protect the frightened tourist, Carey whacked at Rico, but the stick snapped in two. "From that moment," she said, "he knew I was weaker than he was. We look big and strong, but once they find out one of us isn't strong, they don't forget. Rico waited for me for a week until he caught me alone on the dock. Then he attacked."

"You mean he actually remembered the soap incident?"

"Oh, of course! These are primates. They're

very smart and they have excellent memories."

"So it was revenge?"

"It was his strength versus mine. He grabbed my ankles with his feet and one of my hands with one of his and just started biting. Fortunately he didn't have his canines yet. I started yelling. It took about ten minutes for somebody to get down to the dock, but by then he had bitten through the muscles and tendons of my calf, knee, and lower thigh. I can't walk very far. It's a big problem in my work."

"What happened to Rico?"

"Biruté wouldn't do anything with him. Until he bit her. Then he got sent here to Natai Lengkuas — this was before I started working here in 1984. After breaking into the post, he was placed outside the park on the other side of the river and one day he turned up at someone's house and of course terrified the women and children. The men were out fishing and the women jumped into a couple of sampans, which was smart because Rico demolished two houses and all of the food. When the man went to Biruté and said, 'You owe me,' maybe she gave him some money, I don't know. Anyway no one has seen Rico since."

"Justifiable homicide," Trevor muttered.

"I wouldn't be afraid of a wild orangutan," Carey went on. "They don't want anything to do with you. But the rehabs have no fear of people. They're dangerous." We talked on without any sense of rush. Then Carey suddenly stood up.

"We start our days at five A.M.," she said. "It's time to go to bed."

On the boat, in the still of the night, I pulled my cotton blanket from the Kumai market up to my chin and tried to understand why Biruté would place the good of an ex-captive orangutan before the protection of the wild ones, and why Carey was upset about an article that depicted her washing her clothes on a rock.

My feet still itched. Not from the fire ants but from something else. The skin had broken out in welts, even on my legs. "It's the water," Riska announced. Carey had mentioned building boardwalks over the swamp so that visitors could see the forest without getting their feet wet, but a river of acidic tea was flowing under the boat and under the trees, and above the stars were so thick that the sky was only a slight interruption between them.

Suddenly I was laughing into my pillow. Everyone here washes clothes but they do it on docks, not rocks. What if the *d* in Carey's letter to her mother looked like an *r?* Like the other great apes, we gaze into each other's eyes to transmit as well as to gather information. Since before we came down from the trees, the eyes have said it all. But words! Spoken, they are so misunderstood; written, they're positively dangerous. No wonder we depend on waivers to protect ourselves.

River of Precious Stones

This was felt by the Javanese to be a fitting descriptive term to be applied to the many parts of the island which were known for their supplies of gold.

— Victor T. King

Esta was fully geared up by 7:00 A.M. I'd been told by Trevor that she should wear a hat, have all her skin covered except for her face and hands, and wear shoes that covered her ankles. She was going to spend several hours walking through water up to her neck, stumbling over clumps of roots, unable to see through the milky water that harbored snakes and crocodiles.

Trevor and Carey have been planting trees of various kinds as well as measuring the growth of various species. This was to be a measuring trip, and that morning the two of them looked like models from a Banana Republic catalog. Trevor was particularly dashing in black shirt and pants, high black lace-up boots, a machete strapped to his belt, and a black piratical kerchief on his head. He has a short beard and wonderful bones, and all of it made an impression.

I watched him instructing the Trekkers about

their shallow sampans as they lowered them into the river from the high dock and then climbed in gingerly. I watched Esta standing to one side and was reminded of her child-self, silent and concentrated in new situations. As they set off, their canoes dangerously low in the water, the tree-measuring affair did not look promising. Trevor had told me with a guilty smile that crocodiles had their uses — they kept unwanted visitors away. I watched the three boats take off at a snail's pace, each minor wave or ripple causing the passengers to swirl their paddles so as not to capsize.

Once they were out of sight, Kristin, Riska, and I decided to stick with our plans and go have a look at the gold mining that is poisoning the river. Dr. Muin's water samples had indicated that the mercury in the Sekonyer is reaching dangerous levels of toxicity. It is so toxic that the rare and extremely valuable dragon fish Biruté tried to save from poachers a few years back have been eradicated without the poachers' help. Everyone who knows anything about mercury knows perfectly well that the people who live along the river drinking it, bathing in it, and eating its fish and shrimp must, like the dragon fish, be absorbing toxic amounts of mercury, but the damage takes several years to be apparent, so it is easily denied by those who require visible evidence.

As *Yadi* eased the Garuda out into the middle of the river and we began our slow meander, I remembered the speck of envy I had seen in

Carey's eye when I told her we were going to Jemantan. "They'd never let us up there," she said, suggesting they'd be "done for" if they even tried to look around. But now we were leaving the four stations of the park behind and traveling to an area on the other side of the river that has been opened by illegal prospectors, watching the river widen, then narrow, then widen again. Someone was fishing the shallows from a sampan. The driver of a feeble, listing kelotok waved to us as we passed. A kingfisher flew across the unhazy sky.

"Here is where something terrible happened," Riska said, pointing to a place in the water. "Two people were, umm, 'you know,' on a kelotok. They had rented it . . . but . . . anyway, the worst that can happen on a riverboat is sex. It brings such bad luck. If you see little bamboo baskets dangling under the prows, it's probably because of that. Even sexual thoughts on a boat require some offerings."

Around a bend and far upriver, where the lush vegetation suddenly thinned, we saw several boys frozen in place on a white spit of sand as if they'd been turned to salt. Under the intense sky, they seemed as remote from sex or other human enterprise as it was possible to be. They stood on the sandy point at a bend in the river neither speaking nor sitting. When they raised their eyes enough to see us approach, they did not raise cap or hand, but went on staring at the river flowing inexorably down to the Java Sea.

Riska said they were waiting for a water taxi to take them farther upriver. "To Aspai. On the chance that there might be work for them." Java boys, most likely, or Melayu from Tanjung Harapan. On the chance of another shot at digging in the sand for gold, they were willing to stand in the hot sun for hours, without any shelter but thoughts of a better life. There was no magazine or newspaper or cassette player among them. Nothing but time.

Yadi brought the *Garuda* as close to them as he could, but the shallows at the bend made it impossible for the boat to get within fifty feet of shore. We might have waded but there was the mercury to consider, for here and at other mining sites it is dumped directly into the river, gallons of it, after being used to trap tiny droplets of gold on soggy mattresses. Yadi stripped down to his yellow underpants, and we were given the sight of his muscular thighs cutting through the poisoned water. We, who rarely bared our legs or arms for fear of offending and who loved the sight of the beautiful undressed men on the afternoon river, were given this sight of Yadi's flesh wading into the runoff from the jerry-built mines. In a moment, Anang had leaped in, too, despite my shouts and protests. Both boys looked back proudly with great smiles. This sacrifice was for us. They were going to bring back a sampan that resembled an old wooden shoe, a sampan so small and leaky that they could pull only one of us to shore in it at a time. They were risking their health so that we would not risk ours and I wasn't proud of us, at all.

Gold was discovered at Jemantan five years before, and it's mostly panned out there now, with only a hundred or so people left — a few battered huts, a few holes and hoses and gimcracky chutes, and an expanse of glittering white sand where there used to be rainforest. It was rainforest because it had always been rainforest. Now it's sand because it's across from the national park and under completely different jurisdiction. Or none at all. Yes, the mining of gold is illegal, but the trusty park rangers are not authorized to stop anyone outside the park from doing it. Nor can they stop anyone from cutting down trees across the river, although, without a permit, that is illegal as well. Huge flotillas of logs float down the river every day, and no one does anything to stop them.

One of the biggest plywood factories in Indonesia is Pak Aju's, in Pangkalan Bun. Pak Aju, owner of the Blue Kecubung and Biruté's business partner.

Like maidens on the way to our own sacrifice, we were conveyed one by one across the water and assembled on the desolate shore next to the boys who stood waiting for a water taxi to take them upriver. Riska led us up the shallow bank and across a waste of sand to a cluster of huts and working mine shafts. It seemed impossible that only five years before, this white hot sand, burning like a pan under our feet, had been the bed of a forest. In fact, the sands are so white that were it not for the stifling, oppressive heat, we might have been approaching a gleaming hotel with a friendly waiter pointing to a shaded table, a cushioned chair. Perhaps that will be the next order of events, the next incarnation of this ruined land. A sparkling, chlorinated swimming pool. More jobs for the uprooted Javanese, less risk to life and limb. But for the moment, we didn't want our feet to make the slightest contact with the stagnant pools we skirted as we walked through this wasteland.

From around the corner of a shack, a young boy appeared. He looked slightly odd, but what was it about him? He had the look of someone not right in the head, as we used to say. He was leering and bobbing confusedly. Behind him, two men were working in a hole up to their waists in toxic water. At the open front of the shack sat a man who sells groceries from its dark interior, his legs dangling over the threshold, feet not quite touching the ground.

Hajji Makmur came to Jemantan in 1990 with a bucket of rice and a few tins of sardines, an investment of a few rupiahs, but all he had in the world. He sold the rice and sardines to men who traveled up the river to find gold. A little selling, a little buying, a little selling, and now he sits in the doorway of this hut full of produce in what is left of this once-thriving village, a village that sprang up to house the miners who sprang up as well. His bare legs dangling, his sturdy arms pushing against the floorboards so that he leans slightly forward, Hajji Makmur is aggressive in spite of himself, as if he is ready to fend someone off or defend himself from the hot sand that stretches on all sides. His expression is angry by habit, but there is a smile that has the off-centeredness of missing teeth. It's this crooked smile that permits our approach, though the rest of the face is glowering and the arms are now crossed over his belly and chest.

"Ask him if I can take his picture."

Yes. And he would like a copy for himself. With this, he got up and went over to a shelf in the dark rear corner of the hut, took off one *kopiah* and replaced it with another. Black, to indicate his status. He is a man who has been to Mecca. In five years he has sunk three mine shafts and made such a fortune that he has educated his thirteen children and given himself a pilgrimage. Back in the doorway, he sat down again, only this time without even a crooked smile. He sat unblinking and unflinching, as if I were using an ancient camera that required all his patience, and I took a long time with the photograph to reward it. Under the *kopiah,* Hajji Makmur was covered in sand and mud, but his dignity was intact. When another man sat down in the doorway at his side angling to be in the photograph, Hajji Makmur ignored him. Where were we from? he asked.

Canada.

Soon he would go there, too, Hajji Makmur

asserted. "And how old is she?" He pointed at Kristin, asking to have her, please, in the picture with him. Hajji Makmur has had seven wives. Why is she traveling without a husband? And why am I? We had waded through sun and mercury to speak to him. He would show us his gold nuggets.

Around us empty oil drums, filth, and water running through it yellow as piss. With three shafts and his own equipment, Hajji Makmur earns five or six grams of gold a day. This means that ten or twelve grams are flushed out of the water and onto a sticky mattress. The gold is peeled off, the mercury washed away, then he keeps the first two grams to cover the cost of gasoline and splits the rest with his help.

Riska held out her hand and he put an eight-gram nugget into it — a ball the size of a fingernail and quite heavy to hold. His worn face looked cherubic as we passed the nugget around. The effect of a nugget of gold is undeniable.

As we left, we shook hands with our host, the other man in the photograph, and yet another man who had crept up without a word to sit on the doorstep. Hajji Makmur offered me the nugget for the equivalent of forty dollars. Reluctantly, I put it back into his hand. We walked back to the river past tumbled huts, hot, wasted pools of water being pumped and churned or lying spent. At the river's edge the young men were still waiting for a boat to take them somewhere else. A man claims as much land as he wants and holds on to it by working it. That's the unwritten law of any land claim. Usually there is a Javanese company skimming off most of the profit. And when the land is worked out, a man moves on. "Here there is no robbery," Hajji Makmur had said, "but downriver there is. Downriver ten men have been buried alive. They dig too deep in their excitement at finding gold. They are cleaning the sides of the sand pit with water and suddenly the walls collapse and they are thirty meters deep. This was the case in Pangkalan Bun, where the bodies could not be retrieved, but up here we always dig out the bodies. We dig down at the most only seven or eight meters, so this is always possible." He shrugged. "There have been many such deaths."

Holding the thimble-size nugget, I had imagined it stretched and spanning my wrist, thinner than my great-grandmother's bracelet that had been stolen from our house in Toronto. How many nuggets had gone into that lost heirloom, now perhaps melted down again and sold for its value in grams? And the mercury seeping into the river — where there were two women in a sampan fishing for shrimp — how heavy was it? Birds, fish, monkeys, and men slowly absorbing the magical element through what they eat and drink, through their skin when they bathe and their clothes, which are washed in this Sekonyer river of life and death so that Hajji Makmur can go to Mecca and my great-grandmother and I can wear a bangle on our wrists.

Man of the Forest

That evening Esta had just enough life left in her to give us an account of her labors, to which only the labors of Sisyphus could have been compared. "Six hundred trees! Trevor says they've never done that many in one day!" And all of it up to their armpits in swamp and river with huge roots and unseen creatures underfoot and the usual mosquitoes feeding on them mercilessly from overhead. Esta's report included a description of Trevor showing one of the Trekkers how to cut into and suck out a snakebite, after which she decided, as he paddled off to check on his wife, that she felt safer with Trevor and his great machete than with the young Trekker and his penknife.

"Oh any day!" she exhaled.

Measuring the circumference of a tree at a designated place and recording it involves fancy footwork, stamina, paper, pens, and a tape measure clenched between teeth. The trees Esta and the Trekkers were measuring were only two or three feet apart and "we had to climb way up to the metal tags trying to avoid ants and snakes. Sometimes the vines were so thick we had to use calipers instead of a tape." These were real primeval trees, complete with birds and lizards and monkeys, and after climbing up a hummock and working in the heat, "you'd slip back into the freezing water!" She sneezed. "I think I might be getting sick."

More than accidents, I had feared illness among us — anything that would slow us down or increase discomfort. Now, I was hoping she'd have enough of her usual exuberance left for the next day, when we had been promised a walk in the Camp Leakey forest with Pak Akhyar, Biruté's head forester, a man who'd been with her for seventeen years. (Her husband, Pak Bohap, was working the same job when she met and married him.) I was hoping for full participation as well as health. The promise had been made by Riska, who had spoken to Pak Akhyar, but the anticipation was all mine, because he knew each of the orangutans by sight and long association as well as being a real "man of the forest," one who has walked in its damp shade all his life.

We tucked Esta in with all the balm we had to offer and moved downriver for the night, experiencing like old hands the pleasures of unrolling our mattresses and lying side by side under our cotton blankets, while around us the night and its stars and insects provided deeper comfort.

I had been writing in two separate notebooks. One I kept for personal notes and the other for "field notes" about Biruté or the orangutans or trees. I wasn't at all sure where the journey was taking us, or whether I was lastingly interested in anything in the "field." I was interested in writing, in my family, in the life each of us was leading, in the follow as it unrolled and as I curled up in it. That night I wrote: "All very close now in nonverbal ways. Reading *Hidden Force* on dock. Beautiful moon and stars and Anang starting to sing up top. Beer and pleasure."

But the next morning it was clear that Esta was really sick. Riska thought she had picked up Astra's cold, which meant the virus was traveling up the river with us. This is exactly the kind of thing that can endanger both ex-captive and wild orangutans. So, we faced a moral dilemma. We had spent all this money and trouble to come; should Esta be confined to the boat for the next several days? What about the rest of us?

"Just stay three meters away from the animals," Riska said. Then, to be sure we understood, "Ten feet."

It was greedy of us; but at Camp Leakey, we were met by a slender man of almost excruciating shyness who seemed slightly on edge and I was instantly glad we had decided to come. "This is Pak Akhyar," Riska said reverently. "Did you bring the cigarettes?"

I gave him our gift, and he explained his edginess, telling Riska that Kosasih, a large male to whom he is particularly attached, had attacked another ranger the day before, pinning him down by his feet and hands and starting to bite. Kosasih, known as the King, is twenty years old, and Pak Akhyar raised him. It was apparent that he felt particularly responsible for this orang-utan.

When the attack started, a second ranger ran to help, bringing hot water in a kettle. The first ranger escaped, but he was plenty sore and Pak Akhyar was upset. He's a man of gentle manners who looks to be in his forties, with the soft voice, the dismissive posture, and the wiry movements of someone who has spent his life under towering trees. Biruté's staff has varied over the years and was at an all-time low of seven just then, including two or three women who cook for the men, but Pak Akhyar has been an assistant at Camp Leakey since the early days. He and the others are the "workers" Dr. Muin had mentioned, the ones still at Camp Leakey but under the supervision of a woman who no longer has a permit to work there. "What do they do all the time?" I had asked him. "Her research, whatever that is," he had said. Whatever it is, they remain invisible. They have their own quarters — a row of tiny, dark rooms, and for the most part they seem to be in the forest or else asleep.

At Pak Akhyar's suggestion, we decided on a trail not used by tourists, although it was very wet and made walkable only by those unsteady ironwood logs laid down by Rod Brindamour and lashed together by rattan. Pak Akhyar made me a walking stick with his machete when he saw that I was having trouble with my footing. Esta took my camera. Kristin took my hand. My imbalance is famous but I did fine; it was hot but bearable and Pak Akhyar, who speaks a local dialect, talked in a low, constant murmur to Riska, who then reported back to us what he was saying. Usually the reports were about a tree, its age or significance. The plants clinging to it and to each other were noted. Remedies. Usefulness. This leaf for malaria. This tree for water — its long, tubelike branch can be depended on. Serimbang. Remember that in case of thirst. We walked single file and I had trouble hearing Riska unless I was directly behind her, but it seemed to me that we were lucky to hear any of it. "This tree," Pak Akhyar said, pointing to a giant next to me, "is more than one hundred years old." Which means it has survived many slashings and burnings.

Camp Leakey is surrounded by secondary forest that has grown up in the thirty-five years since the area was made a reserve. Before that, some of the soggy land we were walking on was used for rice. There are occasional older and bigger trees, but most of the ironwood is gone, sold for lumber. We stepped over great slippery logs and fallen trees. On one of them where the forest had not yet grown back, Esta found a turtle, but there was almost no visible animal life. Any piece of this forest is rewarding, but one needs the orangutan's gaze. We saw no birds but heard them. We were not bothered by bugs. Pak Akhyar handed a lemon mint to each of us. We trudged on.

What is interesting about walking in the rainforest is that every moment must be attended to. There is no drifting off into unawareness or even deeper consciousness, at least not for the likes of us who grew up with paved streets. This attention must be multiplied a thousandfold for the forest's original inhabitants, since they are concerned with identifying sounds in order to eat and stay out of reach of predators. What I mean is that walking in the rainforest is work. There is balance to be considered and tested and constant choices of wet, slippery, this branch or that. There is the temptation to touch and sniff and inspect — often rewarded by a sting or burn. There is damp and there is astonishing heat, so that we were soaked and panting even in the shade, all the time.

We listened to Pak Akhyar's reedy voice and there were other voices, too. The leaves converting sunlight to sugar. Buds and seeds humming, their juices murmuring. Above, in the realm of the civets and squirrels, fruits and the friends who eat them. Durian, rambutan, mangosteen. Strong-smelling words. There, too, the epiphytes.

This walk could go on forever, given the trees and given the time, and even then we would enter and reenter the dark. Under the trees, the ground would be wet and only a few plants would survive: the palms, the fungi, the parasites. Everything from above would find its way down, falling or gliding, breaking and dying; the beautiful blow of decay would be everlasting and constant. In its damp, in its warmth, constant, too, in the waiting leech. The floor of the forest offers its mold; it offers its flatworms that throw off buds of themselves to multiply and eat. Bellyteeth.

In the beginning, the strangling fig comes to life in the towering arms of a tree, throws down roots that embrace its host, which suffocates underneath. In the beginning, the strangling, the death, the rot. And by the time the host is crumbling, the fig supports itself on holy spirit. The epiphytes live in the canopy, drinking rain and eating leaves. Same beating mouth for sixty million years. In puddles, the tads of treefrogs swim and grow suction feet. Nearby a nest of spiny rattan protects itself from ungulates.

Beetles rolling bits of dung. Caterpillars dressed in poison cloaks. Termites and toadstools and threads of the maiden's veil seizing the day.

Borneo's forests are now only half their original size because the great dipterocarps have become one of the world's major sources of hardwood timber. Only in the most remote regions (where, unfortunately, legal protection is nonexistent) are there still the primeval forests where everything and all of us began. But now, having reached a rainforest intersection — a place where indecipherable paths converge — we stopped to discuss routes while trees reached up and up and flotillas of insects danced in the brief spots of sun. Details. Texture. Degrees of wet, curled, straight. Smells: how they merge and disappear like paths. The choices were an hour to get back through open fields, or two hours on an unused forest track. Pak Akhyar said he hadn't been on the track in many years, but he said it impassively, as he said everything.

I was for adventuring, but there was the matter of time, ours and his. Then he said he didn't know where the track actually led, and that made me vote for it. That and the knowledge that an hour out in the open sun, through those fields once used for rice, would be the death of me.

Unfortunately, we were soon bogged down in mud, vines, fallen logs, morass. This was the forest without Rod Brindamour's trails. In fact, in this part of the forest, there was apparently no trail at all. Doubling back to find another one, we heard the crash of an upper branch. Pak Akhyar, Kristin, and I dashed off in several directions. Esta had stopped. She was sick, about to faint, but she looked up, then called out. "Orangutan!" And this one was entirely different from the ones we had seen in camp. For one thing, she was at the top of a hundred-foot tree, so it was nearly impossible to make her out. But we could tell that she was up there, that she was reddish brown, that she was eating bark. I took several photos, although even with the telephoto lens she was merely brownish and blobbish. When she moved, it became just possible to imagine that she had a face.

Biruté's instructions would have been to stay

with this orangutan for several hours, until she built a nest for the night. This would be the follow. Once she had settled, we would be allowed to go back to the camp to sleep, but we would have to be back at the nest site by dawn in order to keep following. Orangutans settle in a different tree each night. They cover a lot of ground. Or they hardly move. Already my neck ached. My head was tilted completely back, but I could see only a vague shape at the top of a tree.

During a follow we would have to pick up pieces of whatever our subject was eating, so I began stumbling around the base of her tree, filling my pockets with bits of bark and an odd-looking nut she had dropped. Each piece felt like a bit of treasure, something vital to human knowledge of this elusive relative. Perhaps I'd acquired, in those moments, some piece of the forest as yet unknown to be orangutan food. Perhaps, I thought, I should hand my small cache to Pak Akhyar, although he seemed entirely uninterested when I held out my offerings. We watched our orangutan raptly for about twenty minutes, during which time she moved only inches.

When I opened my fists, Pak Akhyar said he didn't recognize the bark or fruit I'd picked up. He said he couldn't tell whether our orangutan was male or female, or how old it was. At that, I wondered about the information filling the field notebooks of eager OFI groups or the young, less-experienced rangers on their own follows.

Still, we had seen a wild orangutan. It must have been wild, since whoever it was showed no interest in us.

We each had a biscuit to celebrate and then walked on through our joy, sweat, and the devastating heat of the old ladang I had hoped to avoid. Here the trees fell back and the path opened and the sun poured itself on us like contempt. Here there was not yet the towering second growth but only scrub and sun and the dry crunch of ferns baking and shriveling, curling their leaves like withered fingertips. Pak Akhyar led us very slowly, speaking now and then in a voice that floated back to us like the call of a reed. Speaking of things here and now, of the trees that are planted in ladangs once the rice is harvested: rubber and palm. Sometimes fruit. The cycle of swidden agriculture I was raised to call "slash and burn."

To establish a ladang, a family must clear an area of ground and burn all the cuttings and stumps at just the right moment before planting so that the necessary nutrients will enrich the soil. If any part of this is done improperly or if the gods and spirits are not kept satisfied, the rice will not flourish. This isn't as detrimental to a forest environment as I was led to believe by my teachers, unless it's done too often by too many people in too small an area, because these ladangs are only used for one or two years. Then they lie fallow and eventually the forest begins to creep back, although *eventual* is the operative word. It won't happen in the slasher or burner's lifetime.

A cleared forest may, when abandoned, regenerate, but its capacity to feed orangutans will not be restored for sixty years, for how can it plant itself, nourish itself, feed the dispersers of seed, without the stored energy of the treetops, without the fruits and nuts dropped to the ground? What grows back is something different, and if the cleared places are large enough, as in lumber clear-cutting, nothing grows back but grasses. There are no other seeds left to disperse.

To me it was familiar — this tall grass, this sward — for I come from a place laid bare. I should have been able to hack it, but it was hotter than hell and I had to keep going, had to move at the pace of Pak Akhyar. Even Riska was trying. To keep up, to walk abreast of our guide, to absorb all the news of this world he had to share. Her delight in the forest is ferocious, for which I already loved her. It was hard to imagine the forest without Riska clambering through it, waving her terrible hat, freeing her long stream of hair.

Finally, we stumbled back into Camp Leakey, its old huts and houses, its green-shuttered rooms. We stumbled back into our lives, stumbled again on a yellow cat dead on the trail. The dead cat upset us, so we showed it to Pak Akhyar. Marched on. To the dining hall. To shade. To a bench. There are so many cats around Camp Leakey; I wondered if they are regarded as personalities to the extent the orangutans are. Just outside the dining hall is the orangutan cemetery, with its poignant wooden markers.

"Maybe it ate a poisonous lizard," someone said, identifying herself as Louise, from the OFI group. She was alone, waiting for the others. This was her fourth visit to Camp Leakey. She's a photographer, and she was taking pictures of the orangutans. "For Biruté. So she'll know how they're doing."

Horrible to think that Biruté was so cast out of Camp Leakey that she could only follow the lives of the orangutans she raised through photographs, but this was apparently the case.

As the afternoon rain began, I descended into gloom. The dead cat growing soggy in the path, a woman exiled, orphaned orangutans . . .

A great desire to move the cat came over me.

But now the rain was coming down in sheets, and we had been joined in the dining hall by our OFI comrades. Kristin stood talking to Eric, from Bolder Adventures, who told her that their research consists of taking notes every five minutes for six and a half hours. These are some of the things their notes are supposed to answer:

1. Which trail are you on?
2. What direction are you going?
3. What is the weather?
4. What are the sounds?
5. What is the "subject" doing?

It was raining furiously.

On the way back to the boat we saw a civet moving across the wet ground by the ramp. After the dead cat, the sight of its wild relative was another reminder of the hazards of domestic life, but we dashed down to the boat through the pouring rain, climbed inside, and drank hot tea as we pushed off from the dock.

I grew up with my brother's collie-shepherd, a woolly, noble cur who flew into my father's lap when he was driving a jeep across an artillery range a year before I was born. Gun-shy she was, hating loud sounds. Maybe even shell-shocked — who knows? But ferocious in her youth and, like most dogs with a shepherding inheritance, a car chaser. Her name was Powder, for obvious reasons, and she was never fully loved by anyone except my brother until her old age, when, deaf and nearly blind, she became suddenly dear to me as a transient being. I took her for long walks, during which my mission was to get us thoroughly lost. Then I depended on her to find our way home.

Maybe for this reason I have loved animals. Maybe for this reason I have never been afraid of being lost. In fact, I remember those wanders with Powder as some of the happiest hours of my life. I treasure them, partly because they tell me something about the young and unchangeable me, something about adventuring and trust and nonchalance and let's go and see.

I am a domesticated hominid. She was a domesticated canine. A long way from progressing to the wild but still harboring some ancient glimmer of it in our more or less cultured brains.

Because I've grown up around them, I'm attached in a visceral way to both *canis* and *catus*. The thought of a wolf or coyote makes me shiver with pleasure. Cats of whatever size or type are beautiful to me. In a kind of serial monogamy, my life is demarcated by individual cats and dogs I have loved. There are people who love primates the same way.

Looking back at Camp Leakey as we puttered off, I did not think I could ever be one of them, and in spite of my wanderlust, I was glad to be more or less dry, folded into the blue wings of the *Garuda*. Then I remembered something.

In the days before we knew we'd have a kelotok full-time, Carey Yeager had reserved two rooms for us at Tanjung Harapan. Now, as the rain came down in sheets, it seemed silly not to use them. At least for a night. The rooms had beds, and there would be four plastic chairs. It would be nice to be under a roof, standing up. Under a roof, walking bipedally. So we decided, on the basis of these luxuries, to spend a night indoors. Only a night. This would give me a chance to talk more with Dr. Muin, who had the third room in the cabin. We could talk about gold and mercury and orangutans. We would invite him to cross the river with us and have dinner at the Rimba Lodge.

The Forest for the Trees

Tanjung Harapan, the first station, sits on the banks of the Sekonyer River where crocodiles swim in the vinegar tea and where mercury, from the mining directly upriver, contaminates everything. Because of pollution, instead of tea, the water looks like milk chocolate. In fact, upriver, at the confluence of the polluted river and the clean branch that leads to Camp Leakey, the streams appear to be coming from opposite ends of the spectrum, one black, one white. Downriver are the nipa palms and mangrove trees. Across is the village of Tanjung Harapan, and the Rimba Lodge. People no longer live in the park, but the signs of human settlement are still evident and will be for another two hundred years, if there is anyone left to notice.

This is where we would spend the night, sleeping in the same house that Gistok and I had investigated when he took me by the hand and brought me across the muddy grass in order to look underneath. He was showing me shelter, I thought; that's what it was. In case I needed a place to sleep!

Now, trying to establish our own personal ranges, we brought our packs and food inside the cabin, chose beds, and enjoyed private ablutions in the *mandi,* where huge drums of the poisonous river water sat side by side in the dark, next to a squat toilet we were cautioned not to use. "It's full," the rangers said. A small pail made it possible to douse ourselves, and we had brought soap in, too. There is no point wondering why we bathed in the water of the Sekonyer. Everyone does. It's the only water there is. In the late heat of the afternoon, while Gistok clambered on the other side of the wall and found a tiny hole through which to peep, the cold water felt as if it had come straight out of a Canadian river, joy and protest indivisible.

Astonished at the simplicity of our good fortune, we put on clean clothes and took Dr. Muin across the river to the Rimba, locking the door behind us. Night comes as suddenly as a drop cloth, under which one can see nothing but the southern stars. We looked back at the unlit cabin. Gistok would soon be finding a place to sleep for the night, but there might be enough time to break a window or pick a lock, rummage through our separate rooms and make off with our precious supplies. Dr. Muin shrugged. He had locked his room, just in case. By consensus Gistok had been declared incorrigible. No choice but to live with him. Yadi fired up the *Garuda*'s engine, and in a minute we had crossed the river to the other side.

Yadi, Anang, and especially Riska had been taking care of us from morning to night. Guides along the river are paid fifteen dollars a day when there is a "guest" to escort, but jobs are scarce, and there are few luxuries in any of their lives. When I invited the whole gang to dinner, they seemed mildly astonished. Perhaps I had broken some code, but we sat down at a big, round table and ordered anything on the menu that anyone desired. I couldn't afford it, but I didn't care. I felt old and worldly. I felt maternal. I had been to Jemantan and survived to tell the tale. I wanted to feed, feed, feed; I wanted to break down the little walls that were so plainly between us. Fish, rice, prawns, eggs. Riska had never tasted beer in her life and took a small sip. "Oh no, oh no," she said, laughing, "this is not for me." Dr. Muin relaxed. He became expansive. Probably the sight of all of us at one table was unusual, but so much the better, I thought, feeling expansive myself.

After dinner, Kristin stopped at the main table to say hello to Giovanna and tell her we'd seen a wild orangutan in the forest with Pak Akhyar. At this, Charlotte, sitting next to Giovanna, interrupted. "That's not what we're paying him his big salary for," she said loudly, "to take *tourists* around the forest!"

Kristin winced, hurt by the tone and by the word *tourist,* though that is certainly what we were. She came back to the table upset, and we all got up and went outside. Feeling bruised, we stood on the deck under the moon. It was pure white in the sky, and the swamp and surrounding trees were full of mutterings and complaints. Riska was furious, but also, I thought, a little afraid. "I'm going in to explain," she said bravely. Yadi and Anang had melted away. Dr. Muin had gone off in the direction of the lobby. But Riska went back to the dining hall and knelt beside Charlotte, actually putting her knees down next to Charlotte's feet. She told her that she'd asked Pak Akhyar to come with us because she respects his knowledge.

"Now you know not to use him again," Charlotte snapped. "He's for OFI use only."

We went back to the *Garuda* and then back to the dorm. As far as I could tell, we were the only tourists to have stayed at Tanjung Harapan in a long time. I suppose this is because the guidebook for the park, written by Biruté and Gary Shapiro, of the OFI, describes the Rimba rooms ($11 to $27) as the only commercial lodging on the Sekonyer and never mentions the rooms across the river ($4).

In our cabin, Dr. Muin sat on one of the plastic chairs, and Kristin and Riska and I took the others. I'd brought a beer back from the restaurant and we shared it, locking the door against Gistok and the mosquitoes, while Esta excused herself. She's always been an early sleeper, while Kristin and I grow alert and garrulous as the night progresses. We told Dr. Muin what Charlotte had said, and to our surprise, he agreed with her. When he said, "You should use the PHPA rangers. That is why they are here," Riska insisted on her right as a guide to seek the best for her guests.

"They don't feel safe to walk with," Riska said, adding again, "I want the best for my guests."

Dr. Muin looked away. He said, "And one of them here doesn't like you."

Riska reddened. "They are all my friends."

I watched the exchange, wondering what was going on and feeling protective of Riska — a sudden urge to bite and scratch. "They will learn, and until they do, it's necessary to depend on them," the doctor concluded, leaning back as if he had proved something, even if it was at Riska's expense. "What about you?" asked the doctor, turning to me.

I said that the PHPA rangers — the ones Trevor calls "Dollar-a-Day-Boys" — don't know enough about the trees and animals.

Kristin tried in her best way to soften things.

"They should get together with the guides and learn about the needs of tourists," I suggested, promising that we would use a park ranger on our next walk and see what happened. But I was upset. Riska was one of my own by then, and I was stinging as if I could feel the slap she'd been dealt, although no doubt it was meant to humiliate her in front of me.

There is a hierarchy of authority on this river. The park is run by the PHPA, which hires young male rangers to live at each of the three stations. At Camp Leakey, there is the overwhelming authority, however mythic, however vanished, of

Biruté and of her workers. The workers still inhabit a long, very visible staff house. Then there are the outside guides — like Riska, like Suwanto — brought in by tourists as personal escorts. Most of these are young men from Java. Riska is the only woman and the only Dayak.

Earlier that afternoon at Camp Leakey, while it was still hot and long before the rain, we had gone down to the dock for lunch and a quick swim in our clothes, with boys from two or three boats doing their laundry and with Yadi diving boldly in to scare off crocodiles. A Dutch couple had arrived with Suwanto, the young man who had wanted to be our guide originally — the one who had rescued us from bedbugs — and Esta had grimaced while he regaled her with a story of the King, Kosasih, examining a female orangutan's genitals the day before and then having sex with her. He seemed to find this amusing. He told me the same story later, but I pointed out (I hoped brilliantly and in order to shut him up) that the female orangutan, whose name is Princess, has a two-week-old baby and therefore Kosasih could not have been consorting with her.

Pak Akhyar is no scientist, but he has lived with the forest all his life. He has learned the trees and tasted their leaves and seeds. He has worked with orangutans long enough to know that, like humans and unlike the other primates, female orangutans may have sex for the fun of it, but never when they're carrying an infant. Adolescent or subadult males are the troublemakers — the ones who rape, as Scott Atkinson had so aptly put it. They follow consorting pairs around waiting for the older male to move away, and sometimes they grab a female, pin her down, and take their pleasure despite her kicking, struggling, and screaming, while with another female they'll show sweet affection, putting an arm around her and kissing her face and ears very gently, as if learning a more delicate skill. Anyway, there was no point in going into all this with Suwanto or Dr. Muin.

In Borneo they say there was a time when all humans spoke the same language. Above loomed the mountain of the gods, reaching into the sky like an ironwood. Sometimes a bear dog came down from that mountain and lifted his leg on the rice spread out to dry on mats in the sun. This was very tiresome, so the humans decided to knock down the mountain. That way the bear dog wouldn't bother them. They gathered together and built three scaffolds to lean against the mountain, but it was hard and their only tools were stone axes. Finally, after they had worked a long time, the color of their skin began to change. Those on the lightest scaffold became white. Those on the ironwood scaffold became brown, and those who built their scaffolds of bamboo were covered in scabies.

Even so, the humans worked at the rock with their stone axes and their scaffolds until the mountain at last began to topple. Two of them were chosen to deliver the final blows while everyone else ran to escape the crush of collapsing stone. As the mountain began to crumble, it crashed into the rivers and plains, and the waters rose, covering houses and fields and forests. The water was so deep that the people ran to untie their canoes, but the canoes were tied to the drowned trees and it took a long time to find them. The humans became hungry while they were searching and looked for something to eat, but all they could find were mushrooms and the mushrooms made them sick. After eating them, they began speaking in different tongues.

I should have reminded Riska and Dr. Muin of this story, but instead I went to bed, stubbornly saying a final "It's not just for the good of the tourists, it's also for the animals," as I shut my flimsy door against the tension in the cabin.

"I've seen the rangers tease them with sticks," I whispered, to myself, because by audibly coming to Riska's aid, I might do more harm than good. The thought hit me forcefully and I remembered thinking the same thing earlier about Biruté and the orangutans. Then I lay in the dark in a wooden box of a room thinking that here, on this river, on the edge of this forest, surely I had not done any real damage. In coming by airplane out of the urban twentieth century and into the forest, surely my own intentions were entirely virtuous.

But what had happened to the people of different colors who spoke in different tongues? When the water level began to fall, the humans untied their canoes, but some were tied with rattan, easily cut, and those humans were swept away so fast that they were carried out to sea and became the ancestors of the white people. Some of the ropes were made of liana, and the people with those ropes were slower to cut their boats loose, and the waters were lower, and they went only as far as the coasts of Borneo, where they became the Muslim ancestors. But those who fastened their canoes with bark rope could not manage to cut them free. They had the onerous task of untying all the knots they had made, and by the time they were finished, the waters had disappeared and they found themselves stranded in the forest.

Wild Thing

Talking to Pak Atak was Dr. Muin's idea. "He's right across the river," he had said the night before. "He was at Camp Leakey for seven years, so he can give you a good idea of what it's like to work in the forest." This is the way my follow went, one person pointing to another, and that one to another. Ripples in a pond would be the obvious metaphor, but I was longing to get to the center, not to the outer edge.

On the other side of the river, word must have gone out that we were coming across. We disembarked at the dock with its oil drums, where there was usually a group of mothers and infants, but this time most of the village was waiting outside Pak Atak's house. The rest were inside.

Pak Atak is a Melayu wood-carver with a strong, inexpressive face. His body is sturdy. His haircut and features are blunt. He makes images of orangutans and proboscis monkeys from ironwood and sells them at the Rimba Lodge, where, except for a couple of baskets and the occasional woven mat, they are the only available local craft. I had thought of buying one of these carvings for Anne Russon, who had been such a help to me in Toronto, although it would be heavy to carry all the way home.

When I mentioned "Dr. Anne," Pak Atak remembered her, which got things off to a pleasant start. We sat down on the wood floor of a two-room Melayu-style house with its now-crowded porch. The house was not surprising, but something about it, something vaguely government-ordered, made me feel depressed. Months later, when I was invited to meet Riska's family, I would see much the same kind of dwelling. A sheet of linoleum put down as North Americans put down a Chinese rug. Two or three rattan chairs. A cabinet.

No one seems to get introduced in Kalimantan, but I decided that the woman holding a child in the darkness behind Pak Atak had to be his wife and probably the old man in the corner was his father. Most were neighbors, of course. About twenty sat with us in a rough circle on the floor, and many more stood outside on the porch, openly curious. But as we talked, I became aware of something heavier than curiosity in the air, as if someone might be spying, although no attempt was made to keep anything we said the least bit private. I noticed the top of a head at the side window, as if someone had crouched out there in order to listen and not be seen by us.

I told Riska to explain that I wanted to interview Pak Atak, and as she did so, I got out a notebook and pen. At that, everyone fell silent and I asked my first questions.

"When did you work with Mrs. Biruté in the park?" I said. "When did you stop? What was your salary while you were working with her?

Pak Atak said that he had worked for Biruté from 1984 to 1991. His starting pay was 90,000 rupiahs a month (about $45), with a raise of 500 rupiahs each year. He worked as a data collector, starting out at 4:00 A.M. on a follow and keeping track of the time his subject woke up, what he or she ate, and which trail he or she was closest to. He told me it was Biruté's practice to follow the rangers out at nine o'clock. There were seventeen data collectors in those days. None of them had been trained.

"There have not been any bad incidents," Pak Atak said, and with that, he opened his hands. "A woman was almost raped by Rombe, a subadult male. An adult female called Supinah doesn't like women to walk with men. Six women got bites from her. She is now wild."

"How many have gone back to the forest?" I asked.

"Maybe sixty." Pak Atak seemed eager to talk in spite of the audience sitting on the floor under pictures of beauty queens saved from a calendar. The cabinet stood in a corner. It had glass doors and held both food and dishes. When I asked a question, Riska translated. Then Pak Atak answered, and sometimes one of his neighbors as well.

When Pak Atak quit working at Camp Leakey, it was because he wanted to earn more money, he said. It was probably about the time he met the woman who was sitting slightly apart from us, holding their child. He went to work for Dr. Carey in 1993, but still earned only 90,000 rupiahs a month starting pay and was still dissatisfied.

Because he worked at Camp Leakey during the time Biruté and her staff took on the care of the Bangkok Six, I asked what had happened to them.

Pak Atak hesitated for the first time, and someone said, "They all died. In Tanjung Harapan. From the water."

"All of them?"

"Yes."

"Was this upsetting?"

Same voice. "No. They aren't human. But Mrs. Biruté cried."

"What was wrong with the water?"

General chorus. "They use medicine for the ramin wood. The cutters. To season it. So it doesn't rot. The medicine is poison. Now PHPA made them stop, but the ramin is all gone anyway and the other trees don't need the medicine."

"When did Mrs. Biruté stop spending so much time at Camp Leakey?"

"Around 1990," Pak Atak answered.

"Why was that?"

"Maybe they are bored because they have families. Mrs. Charlotte got married."

"What about having tourists here? Is it a good idea?"

"It is a very hard problem. More ex-captives are getting tame because of more tourists. They are getting tamer, not wilder. But they need money to support the national park. Mrs. Biruté thinks it disturbs research because tourists want to hold orangutans. Mrs. Biruté says if tourists come, okay, but they should make data. Write everything down."

"When did you start carving?"

"In 1993."

After I had looked at the two pieces Pak Atak was working on, I asked if he could finish one by the next morning. I told him I wanted to take it back to Dr. Anne and he smiled for the first time. He said she had wanted one of his carved orangutans, but he hadn't had one ready and now she didn't come back to Camp Leakey. Anyway, she never got her carved orangutan. We negotiated a price. The carving was heavy, but Esta generously offered to carry it home in her pack. She was leaving Indonesia a week earlier than Kristin and I were, and flying direct.

At last we stood up, stretched, thanked the family, and backed out of the tiny house and through the throng of listeners on the porch, arranging to come back the next morning for the carving. I was glad to have the right present for Anne and glad to soften the hardship of our visit to Pak Atak's, which felt like an imposition. We walked around the village, most of which was a foot or so underwater, stepping carefully along boards laid down in the mud. I took pictures of the teachers' house — three small attached rooms, three separate doors — a village dog, a group of children standing just out of the water. One of the children was wearing enormous sunglasses. Behind him a woman sat on the porch of a waterlogged house with more children. The water in her yard gave her small house the look of yet another ark, or had that image become fixed in my brain? This was the dry season, but only the planks of wood made walking possible. Children. Dogs. Roosters. All seemed dull and dismal.

Later, docked at the Rimba Lodge, we borrowed two sampans, one of which I instantly capsized, to the amusement of everyone standing around. Our volunteer paddler flew into the river, soaking himself and, more important, his cigarettes, which I later paid for. He was not amused. Feeling large and awkward again, I clung to the sides and sat very still as he paddled.

To avoid being swamped by the current, we went along the edge of the river, winding and bumping under overhanging boughs of pandanus, Kristin and our paddler fore and aft in my canoe, and Esta and Riska in the other.

The river is bigger when you are down in it.

In a sampan, the waves are immense.

Up a tributary, then, and through swamp until we emerged behind Tanjung Harapan village right in the middle of their ladang. Because of the swamp, these villagers had to learn how to plant wet-padi rice after moving here from the other side of the river, where the land is dry. They didn't want to move, but Biruté insisted, so the government moved them and built them new houses to compensate them for the loss of their burial ground. Biruté apparently felt there would be continued poaching and tree-cutting in the park if it had to support human beings. She wanted clear boundaries. So the villagers left a field of small graves behind, marked, in Muslim fashion, by carved posts at the head and foot, like sunken beds set down for buried sleepers. It is a field of blue wooden cribs, a nursery of children whose parents have been moved across the water.

We turned carefully and paddled back toward the trees. From the river, the forest seems to stretch back on both sides, but now it was obvious that the trees on the side that isn't park were simply a screen, behind which lie swamp and ladangs. Without protection, the trees closest to water get cut down first, as they are easier to haul out and float away to a mill. In fact, the screen of trees reminded me of highways in Canada, which are often similarly deceptive, their forested sides lulling the traveler into thinking the woods are pristine. I could almost hear Biruté's voice: *"And the situation in British Columbia is frighteningly similar to what I see in Borneo."*

Then it was back into the dark and down a tributary with a huge white butterfly floating past like a piece of Kleenex, the light of the forest thrown off surfaces of bark and leaf, filtered through leaves so dense that there is hardly enough of it left to allow photosynthesis to take place. Out on the river again, where the smallest waves could capsize us, we had to hurry our sampans to the shelter of reeds when even a small kelotok came by — a taste of what people here experienced whenever we passed.

I saw Esta signal from her sampan. On the park side, close to the water, she'd spotted a young orangutan in low-bending pandanus! This was an unexpected thrill. We paddled up close enough to see him quite clearly but kept enough distance to protect him from us and us from him. After awhile, he made a small gesture, moving ever so slightly closer. "He's wild," Riska said under her breath, adding that it's unusual to see a wild one so close. It was possible, of course, that he had left one of the stations, that he was exactly the orangutan Biruté and all of us were hoping for — a successful rehabilitant — but we were an hour downriver from the second station, a long way for this five- or six-year-old boy to travel. As he watched us, I rustled a bag to see if he knew about packages. Any orangutan who has lived around people will come to the sound of noodles or crackers being unwrapped, but this one showed no interest. So close to water, so unafraid, who was he?

We went upriver and then came back to him, unable to unfasten ourselves from his gaze. I almost hoped that he *had* been rehabilitated, even though discovering him would not be such a thing to write home about if he weren't wild. The prospect of an animal born of a wild mother and now returned to the wild so nicely by its mother's murderers put our species into softer perspective. We looked at him and smiled and liked ourselves a little better than we had before.

That night, from my room, I heard one of my daughters saying, "I'm tired of all the comparing that goes on in this family . . . who got the best piece of batik . . . who walked the farthest . . . who . . ." and I froze in my bed, for it was I who often compared them, wasn't it? My fault, my fault. It was hard not to intervene, almost impossible not to rush into the room they shared, but I didn't. I thought, Unhappiness itself is the imperfection, a fly in the ointment of my idealized picture of us. "My children are my jewels," as my mother and the fairy-tale Cordelia, asked to show off her gems, used to say. "Look at them." Children as mirrors. Children as proof of value. That's probably natural. So is the wish of daughters not to be like their mothers. Then I thought, Mothers interpret daughters. To themselves. To the world. But the task of the daughter is to interpret herself.

My daughters are enormously fun and good and kind to me, and their smallest criticism cuts me to the quick. They were upset that I was taking notes during the interview with Pak Atak, but I had thought it through and decided to do it visibly so that it would be clear to everyone that it was a conversation that might be made public. I thought the presence of paper and pen would convince Pak Atak that his words might have consequence. Still, the criticism by my children made me doubt myself, and perhaps my comparisons (with their implied criticisms) had the same effect.

The more you love someone, the more you interfere.

I stared up at the roof, over which Gistok would soon be trampling. All in all I'd been happy on this journey and intrigued by it. But if I preferred my children's pleasure and happiness to my own, or if I confused it with my own, I couldn't take care of myself, which was probably worse for them than for me. For daughters, maternal passivity sets a bad example. In the forest, a passive mother would be at the bottom of the social hierarchy, and her offspring would inherit that position. Maybe it's the same with *Homo sapiens*.

The conversation on the other side of the wall was softening. Years before, we had lived in a house with walls as thin as these. Then I had found it comforting to have my daughters close. Night hanging around us. Only a rustle of worry outside in the trees.

Riska was in the bed across from me. I closed my eyes, closed my ears, fell into a restless sleep under the aluminum roof, with mosquitoes aplenty and no breeze, because all the windows of the cabin are permanently boarded up against Gistok. Hours later, I woke to his scrambles across the outer wall as five long fingers poked through the unscreened opening between window and frame and a flattened face looking like longing itself pressed up against the smudgy glass. Dr. Muin had gone and the others were already awake. We stumbled to the outhouse, to a breakfast already cooked by Riska, to the *Garuda* and back to the village across the river to pick up a more compliant, carved orangutan.

Pak Atak was squatting on his porch, working away with a small blade. While we stood watching, examining two partially finished pieces he had set out on the porch railing, there was a rumble behind us on the path, and suddenly Charlotte came charging up followed by a nanny holding her youngest child and by an enormous Dayak who looked murderous and unintelligent and is known as Biruté's bodyguard. Charlotte was bellowing and Pak Atak's neighbors once again began pouring out of their houses to see what was going on, but since she was bellowing in Indonesian, I had no idea what she was upset about. All I could think was that she had found me in the act of buying a carving outside of the Rimba's shop, so I tucked the piece I was holding under my shirt. When I tore my eyes away from Charlotte's pulsing neck, I saw that Pak Atak was squatting on his porch exactly as he'd been squatting when we arrived, although he was no longer chiseling. Yadi was leaning against a tree, and a real crowd was gathering, like clouds in a sunny sky. Then Esta broke into Charlotte's diatribe. "Would you mind speaking in English so that we can know what you're saying?"

Charlotte spun around. "Oh! I'm sorry. I'm glad to find you all here. I thought you'd left. That's what I heard. But this is better because we can get to the bottom of it. I know you were here talking to Pak Atak yesterday. I heard about it from a source. I won't say who, but it's someone I can't trace at the moment, and he said Mr. Atak told you all kinds of things about his salary at Camp Leakey, but worse than that, he said the Bangkok Six were all dead, and I couldn't believe he'd do that to me. Of course I wouldn't be able to buy his statues anymore, you can see what I mean, if he said something like that, so is that what he said?"

The bodyguard stood by her, glowering.

"You wouldn't buy his statues if he said that?"

"You can see why."

Esta put in, "But it wasn't Pak Atak. Everyone was talking at once. Riska was translating. It was someone else."

Everyone who had been there knew exactly who had said that the Bangkok Six were all dead, and it hadn't been Mr. Atak, although he had not disputed the information. It had seemed at the time that everyone in the room agreed on the fact. It was their unanimity that had surprised me.

"I see Mr. Yadi is here," said Charlotte. "If he tells me Mr. Atak didn't say it, I'll believe him. He has always been completely trustworthy."

We scraped our toes in the dirt and looked sideways at each other. I held the carving under my shirt. "Didn't say it," he muttered, and Charlotte closed her mouth. Pak Atak had worked for Birutė for seven years, and he was now managing to make a living from his carvings, but only because he sold them at the lodge. In fact, there was no other place to sell them. In one stroke Charlotte had been willing to wipe out this young man's livelihood because of something she'd heard. And she was still ranting, insisting now that the Bangkok Six had been in such terrible condition when they arrived that there was never any real hope for them.

"Two of them died in my own arms!" she cried. "Their lungs were all ruined, that's what the vet told us. He didn't think they could survive."

Strangely, she was not disputing the truth of what we'd been told in Pak Atak's house either, although the OFI has consistently denied that six orangutans died. What she was disputing was Pak Atak's right to speak. "I hate disloyalty," she explained, when her rage had finally begun to subside. I remembered a story I had heard about Birutė, who was supposed to have said, "I should do what Jane Goodall does. She makes everybody who works with her sign a loyalty oath."

"And to think I gave him a watch from the States only two days ago. I just couldn't get over it. I couldn't sleep all night. I had a terrible stomachache," Charlotte complained.

Like bacteria, the effect of such rage can only spread. If I, who was leaving the river the next day, felt afraid of her, the others, who live there and depend on Mrs. Charlotte and Mrs. Biruté for protection and livelihood, must have been terrified. Everyone on the river seems to depend on them in one way or another. For these people, the two white women can make life itself insupportable. There had been the scene about Pak Akhyar, and who knows what repercussions there. At the very least, a stern reprimand and a reminder of his dependence. The whole business of arriving with a six-and-a-half-foot bodyguard and screaming in public is astonishingly out of tune with Indonesian ways. "I'm sorry but I'm American. I'm a Westerner," Charlotte said several times, by way of explaining herself. "This is how we do things. We don't sit around and wonder; we confront!"

"Is that true what she said about Westerners?" Riska asked me later.

I told her that it was absolutely not.

"Because here we never never show our feelings that way. It is the worst thing you can do."

I said I thought Pak Atak had been very brave. He had not spoken up to deny her charges or laid the blame where it properly belonged. He must have been frightened, but he seemed outwardly unruffled. Of course, I may have been seeing things from my own Western perspective. Perhaps Pak Atak viewed Charlotte's display of anger with nothing but contempt. Perhaps he had never felt a shred of fear.

Before Charlotte retreated I had mentioned that I was leaving the next day early in the morning, but that I would like a chance to talk to her. "Good," she said. "I'll be at the Rimba at four forty-five. How's that for you?"

I told her it was perfect, that I was going back upriver to Camp Leakey, but that we should be back by then. "I'll meet you in the restaurant?"

"I'll be there."

We went up to Camp Leakey and had a walk with a PHPA ranger, as I had promised Dr. Muin we would do. The walk took place after a hot, thirty-minute wait for a ranger to be available because, although there were two of them on duty, one had to stay in the office so the orangutans would not break in, and the other had another batch of tourists who wanted a walk in the forest. Even so, once we got started, the walk was good. This time I felt more confident, horribly hot but not miserable, and the forest felt more familiar. Our ranger seemed to talk a lot, but Riska translated less. Either she didn't respect his information, or perhaps they were just gossiping about mutual friends.

We did not see any orangutans.

After three and a half hours of mud and water and fallen limbs, we met Pak Akhyar at a fork in the path, and I could not help being glad to see him. He'd come out to warn us that Kosasih was on the trail, but that wasn't all. Yayat — the other adult male — was nearby. These two orangutans don't mix. If one is in camp, the other one stays away. Primate males of whatever species are hierarchical. No wonder our ranger looked upset. Either he was irritated by Pak Akhyar's interference, or he didn't want to take us past Kosasih in a malevolent mood.

Once again, Pak Akhyar led us back to camp.

I paid our ranger 2,500 rupiahs (about $1), which was less than I had paid Pak Akhyar and probably not enough, but I told myself that he worked for the park, he was employed by the government, it was part of his job. And I was testing Dr. Muin's thesis. As the park closed, we went back to the boat and had a beautiful trip downriver, looking back for a few long minutes at Camp Leakey and already feeling nostalgic. Overhead the trees were full of proboscis gathering for the night, each tree apparently home to a male and his offspring and wives. All that was left to see was a hornbill. And then one flew past. What a sight! The bright, abundant bill was visible from the boat, and I wanted to think that this was auspicious, because the hornbill is king of the sky.

That afternoon there were other signs as well: eagles, macaques. And we got to the Rimba on time. But Charlotte was bathing, according to a man at the reception desk, so I waited for her in the restaurant, wondering what I was going to say. I have come to this place . . .

I have become in this place . . . somehow troubling . . .

Poor Biruté. In an effort to enhance the viability of Camp Leakey, she'd built this lodge in 1990 and expanded it to house the 1991 International Great Ape Conference at which, she must have hoped, reintroduction of ex-captive orangutans would be endorsed by the "experts" and her own reputation, which was suffering, would be redeemed. Her reputation was suffering because of the disaster of the Bangkok Six, and because of a second international smuggling incident, the case of the Taiwan Ten. What happened following the conference, however, was bad news for Biruté. Despite her vigorous lobbying, the directorate general of the PHPA in Jakarta issued a strong statement of its new policy on rehabilitation. Confiscated orangutans were to be released henceforth in "suitable habitat in which no wild conspecifics occur . . ."

That meant that Camp Leakey, and perhaps even the whole park of Tanjung Puting, were no longer viable for the rehabilitation of ex-captive orangutans; too many wild orangutans with whom to mingle.

The most damning part of the statement was outlined by an ominous black box:

> The still continuing rehabilitation programmes of the 1970s have become outdated and cannot anymore be considered as a feasible approach to tackle a problem which now exceeds beyond the illegal capture and trade of orangutans only. As a matter of fact the stations have become a liability for orangutan conservation while they pose a serious threat to the wild population of apes in their surroundings . . .

When Charlotte came in, she sat down at the table and I told her why I was in Kalimantan. "I'm working on a book," I said. "It's kind of about Biruté. About the effects of her work . . ." I said that I had met her in Los Angeles and we had talked, but that I hadn't heard from her again. "Your friend is elusive," I said. We sat across from each other like two women in any restaurant, which is what we were, except that she owned the restaurant and it was surrounded by nothing but river and forest, and the morning's rage was still out there somewhere, palpable and dense.

She ordered two cups of tea.

I said I'd thought about talking to her earlier but she was always surrounded and busy. I didn't tell her that the real reason was something else, something harder to explain. Not knowing how Biruté felt about me, it had seemed unfair to talk to her best friend. But it was tempting. I'd heard that Charlotte disliked the influence L.A. people and the OFI had over Biruté, and I could obviously sympathize. I thought of relating the incident in the parking lot with Nancy Briggs. But after the scene in the village that morning, I couldn't bring myself to such an obvious device.

"But she's not here! How can you write about her?" Charlotte said, and I knew what she was going to say next. "Everything's in her book. What more is there to say?"

"We have a different focus," I offered. Then: "What I wanted to talk to you about is your project in Hawaii."

At the mention of that, she brightened. "Well, if you're interested . . . you should come out some time. Ask for Charlotte Grimm, just like the fairy tales; that's my name."

"I'd like to do that. I used to live there, actually, which is why I find the idea so surprising. Because it's impossible to bring any animal into Hawaii without a four-month quarantine. They're paralyzed about rabies over there. Unless things have changed. Have they? I had to leave my cats with my mother when I went" — I was rambling wildly — "and the ecosystem there, being an island, is so fragile. They brought in the mongoose and look what happened." The whole idea seemed astoundingly out of whack. "How did you talk them into orangutans?"

Charlotte said, "Well, I know someone — in the government — a state senator, and I've got him onside. Name of Andrew Levin. It's been a lot of work, but the climate's perfect, and they'll be in cages. Very large cages. So they can't come in contact with anything." Our tea arrived. Charlotte passed me a cup. "And about this morning," she said, pulling in her chair, "I think I've got down to the bottom of it. I mean, it's all a family thing, like everything is here. Everything here comes down to that. See, Pak Atak was married to a woman and left her for the other one he's with now, and it was someone in his wife's family, the first wife, who spoke to me. No doubt looking for revenge."

With that, our conversation was almost done. I told her again that Pak Atak had told me nothing surprising. I offered to pay for the tea. I might have asked her who had actually *cared* for the Bangkok Six, but I thought I knew the answer, so it never occurred to me. I might have asked what had happened to Biruté since then; had she changed? But I held back. The trees that come down to the banks of the Sekonyer River are confining, but this is a tropical river, an equatorial waterway. I shouldn't have been surprised by the sense of claustrophobia, perhaps. But I was surprised anyway.

Leaving the shiny Rimba Lodge, I remembered when my ex and I inherited, from his grandfather, the huge sum of $5,000. At that time, and throughout our married life, we lived on $300 a month from his trust fund and whatever I could make in wages from whatever job I could find. We had two children. We went to see an investment person — someone who had been recommended as honest — and explained our many moral qualifications for any investment he might make on our behalf. Absolutely no chemicals. (This was the 1960s.) Nothing that might further the Vietnam War effort or improve the profits of anyone who did or might or wished to. Nothing damaging to the earth. (I had read Rachel Carson.) Nothing that profited from injustice or misery.

We did not want much profit, we insisted, only a little wee bit.

The young (but not so young) and slightly slick broker or investment counselor sized us up and suggested we invest in tourism. We stared. We liked to travel, he rightly pointed out, and so did everyone else, especially the elderly, who had been saving up over many rainy days to come to the islands — some of them waiting all their hard-working lives — and now there was a company established to further their cause . . . the cause of geriatric tourism. I remember the nice ring of that.

This was years before it was shown that all tourism in Hawaii pollutes the sea, wastes the fresh water, and leaches the land; that very little if any of the enormous profit from that tourism is returned to the local community, and that local lives are made not more happy but surely less happy as a result of it. This was before that sad news, and we were instantly hooked. Ah, to be benefactors and turn a little profit at the same time!

The point of the story isn't that we lost all $5,000 within mere weeks of signing it over. Or that the young, but not so young, slick broker left the islands without a trace. The point is that the idea of tourism for the elderly sounded so nice and comforting. And we were being good, for we preferred less advantage to ourselves in return for less damage to the planet.

There is a lot to be said for disinterest, but it's hard to invest with it.

And it's hard to expect it of others when our own self-interest is so obvious in most tourism schemes.

Of course, once a national park is created, a tiny local marketplace grows up around it. There will be a need for guides. Postcards. Perhaps some crafts. Baskets made of local grasses. Wood carvings. A few uniforms will be purchased and starched and ironed, and the people wearing them will have real jobs with real benefits. There will be respectability involved all round and maybe that will make up for whatever gets lost in the process. (In the new strategy outlined by the Minister of Forestry's report, another threat in the guise of a promise reads: "It is considered that after due monitoring and research a special sector of habitat can be utilised for guided tours ['ecotourism'] to see re-introduced apes. This shall be developed not earlier than three years after the group of apes in that sector has been feralised and no individual can be expected to seek contact with visitors.")

The hotels and lodges will almost always be owned by outsiders, often foreign, which means the profits won't even stay within national boundaries, much less local ones. The travel agencies, airlines, tour companies, buses, and chartered equipment will likewise be owned by outsiders. Who else has capital for these things? And we know, by all the principles of capital and investment, that he who invests has the right to profit. The rest of us work for a wage.

What happens to the rage or disappointment of people who have had their hunting and farmland taken away in order to secure the borders of a refuge or park? They are promised wealth, at least relative wealth, but what happens when the promises fall short? Now that a huge portion of workable land is gone, there will have to be other alternatives, and even without fanciful imaginations it is possible to think of several. This is, of course, after the animals and trees and rocks and sacred places have been commodified by others into something unrecognizable by those who know them best.

I had gone to bed two nights before feeling innocent. But it was the next day, and the next, and I could see that every move I made on the Sekonyer would have repercussions. I would be gone. The people who were likely to be hurt might even be unknown to me. My journey was my own, but I would leave traces of myself behind. Like mercury.

Gistok Holds the Key

"Speak and I shall baptize thee."
— the Bishop of Polignac

Early the next morning, Esta disappeared. We'd stopped to say good-bye to the rangers, to Gistok, and to the Trekkers, who were working on the new clinic. After an hour of this we'd gone back to the boat. Most of us had. But Esta had not. Perhaps she was refusing a world of cars and airplanes and tight schedules, and who could blame her? So, having said good-bye to everyone, we sat with our legs stretched out on the deck of the *Garuda* and looked up at the trees and across the river at a woman fishing in a sampan and wondered why Gistok had not come down to see us off. He always loped down to any boat that tied up, eternally curious, maybe even optimistic, although his expression was invariably petulant. I suffered my own moment of temptation — a desire to head for the trees behind the little hut. Maybe I would find him high in the branches of the forest, where he belonged.

I also wondered whether I should check on Esta but decided to leave her alone. One thing we had not had a minute of on this trip was privacy, and all three of us require a lot of it. Yadi sat at the wheel and puttered with this and that. Anang was nowhere to be seen but probably only inches away. When Esta came running across the open yard in front of the rangers' hut with Gistok in fast pursuit, things seemed to be back to normal. Except that she was obviously furious. Climbing aboard and pushing off, she told us that Gistok had followed her all the way to the outhouse, where he had demanded to come inside with her, wedging himself in the doorway until she had forcibly shut him out. But Gistok, man of the forest, had his revenge. He snapped the latch shut on the outside and very assuredly locked her in.

Orangutans are not famous for their use of tools, the way chimpanzees are, but they are Houdinis where any kind of bolt is concerned.

A wild orangutan can work himself out of almost any confinement, given an item or two to use as pick or pry, and an ex-captive is just that, an *ex*-captive. He wants control of all keys. He wants the run of things. Gistok was obsessed with interiors. Ho, the city life for me! he seemed to be thinking, as he stared in at our cooking and bathing paraphernalia, longing to rehearse for the role of jester in our endless comedy.

After half an hour of pleading on Esta's part — "Gistok, come on, I'm sorry. Please, Gistok!" — he had finally unbolted the outhouse latch as calmly as he had bolted it.

Of course we all howled with laughter, but

Gistok's escapade got me thinking about him in a new way. When Darwin theorized an origin for *Homo sapiens* in Africa, it was assumed that something had set us *apart* from the apes. It was assumed that the something was language. A big part of the deliberations of the Enlightenment had to do with this distinctive human attribute. But what was its cause? It was thought that language might be accounted for by our "loss of nature," but where is the line between language and unlanguage, between nature and unnature? Language requires deviousness, for example, but deviousness exists, like the love of durian, in both human and ape. Gistok is devious, yes, and, more than that, Gistok is manipulative. It's an interesting word, *manipulate*. We use the same verb for the way we work with mechanical objects and the way we control another person's reactions. Orangutans are devious, but it isn't just tricks that they play. Like us, they conceptualize.

I've seen a zoo orangutan play with fabric, the way a child does, putting it over his head and pretending to be invisible. This is devious. This is ape-type intelligence. Second-order intentionality. Cause and effect. The zoo orangutan is using her intelligence exactly as we use ours. She is *taking our point of view*. Perhaps only great apes make the leap into another mind. We work from the beliefs and intentions of others, and this is part of what communication means. Communication is an act of mutual intentionality.

While formation of words requires motor control from the brain, a particular musculature of the larynx and tongue, and highly controlled breathing, our ape brain provides us with an ability that is essential if we are to speak. We understand intent. We also understand that a word or sound can stand for a thing. We can *represent.*

We can generalize backward and ascribe this ability to whatever ancestor all of us great apes had in common. She must have been large-brained and fast-learning. And there must have been social pressures that caused her to become this way because neocortex expansion parallels social complexity. It used to be thought that group hunting taught us to cooperate and that the need to communicate during cooperative hunts led us to speech. But since the ability to *represent* happened to orangutans as well as gorillas and chimpanzees and bonobos, it wasn't hunting and it wasn't cooperating that caused primates to think symbolically. Something else made our society more complex than that of other species, something that especially affected the upright, walking ape.

Whatever that evolutionary chain of events was, it gave us the ability to understand patterns and to think in sequence. And those abilities allowed us to develop syntax *and* to play tricks. Deception, in fact, is an indicator of language ability since *mis*representation requires the ability to represent. We can play tricks because we understand how another mind thinks.

Maybe Enlightenment thinkers created a false barrier. Maybe language is not distinctively human. Orangutans and chimpanzees have been taught to use sign language and have even passed that language on to their own offspring, making it part of their "culture." When asked questions, they answer thoughtfully. They think, then translate their thinking into language. In 1925, the Yale primatologist Robert Yerkes had written, in a book called *Almost Human*: "I am inclined to conclude from the various evidences that the great apes have plenty to talk about, but no gift for the use of sounds to represent individual, as contrasted with racial, feelings or ideas. Perhaps they can be taught to use their fingers, somewhat as does the deaf and dumb person, and thus helped to acquire a simple, non-vocal, 'sign language.' "

He was right. But Lyn Miles, who later taught sign language to an orangutan named Chantek at the institute Yerkes established, believes that learning to communicate depends enormously on motivation. Chantek learned sign language exactly the same way a child learns to sign or to speak. Easiest were signs for the things he wanted — certain foods, certain responses; hardest were signs for things he had no interest in, like shoes. Down was a word he wouldn't learn, but Chantek learned to sign up almost immediately, because he loved to be carried. Hardest of all were signs for things that weren't present, or for ideas, but eventually Chantek even applied signs for dirty and bad to his own actions, and by the time he was eight years old, he had invented several signs. One was no-teeth, to indicate that he wouldn't bite during play. Another was eye-drink, for contact lens solution. Then he began using signs in novel combinations to express various meanings. Chantek was able to take the perspective of the other and understood that signs were abstract representations. He even engaged in linguistic deceptions, and referred to things that weren't present. Most surprising of all, he began to talk to his toys, investing them with powers of understanding. Ape animism. Ape belief.

Chantek even has me wondering about my theory that only humans save, since, with Lyn, he was paid in coins for making his bed, picking up his toys, and cleaning his toilet. After learning the signs for several flavors of ice cream and, after saving enough money, he was allowed to spend it at the ice cream "store" by "signing" his order. He preferred nutty ice cream, especially pistachio, and sometimes picked the nuts out and lined them up very carefully. Later he ate them, in order, one by one, or saved them for another time.

"He hasn't made a stone tool," Lyn told me when I met her during a visit to Toronto, "but he's close to it. He understands the process of chipping flints and could learn to do it, now, without much difficulty. What he does do is go through a very complicated series of interactions, picking up one thing to get another thing that opens something else and so forth until the ultimate reward is obtained."

As we set off downriver, I was almost sorry to leave Gistok behind. No wonder people like to adopt orangutans and take them home. They are so like us, and they don't talk back. At least, not verbally. We all waved. Good-bye, Gistok. The sight of him standing there, so desolate, on the dock, was the last one we would share of that supervised Eden. Topside, Kristin's hair blowing in front and Esta and Riska stretched out beside me. It was beautiful. It was perfect. But the magical time with my children was almost over. Good-bye to all that, as well. We were going to fly to Jakarta, spend a night there, and then put Esta on a plane back to Canada. Kristin and I would go on, spending a week in Sumatra and having a look at the efforts to rehabilitate ex-captive orangutans there, but that was different. That was not the three of us, as we had once been and would maybe never be again.

The motor blew its breath out against the palms and mangrove trees, and I tried to console myself. During our dinner at the Rimba Lodge, Dr. Muin had promised to give me some PHPA reports that had to do with the orangutans, their ages, dates of acquisition, past histories, and deaths. These would help me understand how many animals were actually resident and how many had gone back to the wild. Since there were three stations and constant changes in two of them, enrollment was hard to track down. There was Winny, for example, a three-year-old female who had appeared during our farewell visit to Tanjung Harapan, although she must have been somewhere nearby all the other times we had been there. Winny has only one arm because when her mother was shot out of a tree, she fell to the ground, and her arm was so badly broken that it had to be amputated.

This horror must have left her psychologically damaged as well as physically handicapped, so the probability of her eventual survival in the wild seemed small. A one-armed orangutan would be incapacitated in the forest, since the usual method of locomotion is clambering through the branches of the canopy, but Winny was able to get around bipedally, since she lived among humans. Ex-captives are unusual in this: they tend to walk on the ground to get from place to place, and their bow-legged, stooped walk gives no indication of their true arboreal grace. In the forest, however, except for the full-grown males, life on the ground is simply too dangerous, even if enough food could be found there. A wild pig would have Winny for lunch unless she could climb very quickly out of his way.

Winny had the size and winsomeness of a year-old human child who wants to be held and carried, who rides nicely on the hip and who clings tightly with her legs while looking around, never missing a thing. I couldn't resist her. In spite of everything I knew about contact, I'd swept her up in my arms and carried her around for that last hour before we left the station, feeling the remembered weight of Kristin on my right hip. The Trekkers had arrived and were hammering and sawing and sweating and listening to a boombox that seemed to have endless battery power. They didn't have time for Winny, but with her one hand she nonetheless tugged at their pant legs and sleeves and managed to be continually in the way. "Oh, Winny, get on with you," they'd say, as if she were a pestering little sister, someone they'd had to put up with all their lives.

Then there was Melly, the first orangutan we'd met, and her year-old baby. And there was Davida, who had started ex-captive life at Camp Leakey long ago and was now fifteen. In Manhattan Beach I'd met the man who had originally taken care of her when she was an infant. He was hoping to come back for a visit since Davida had recently given birth to a baby, but just before we arrived, Davida's new baby had died, and when we saw her, usually sitting in a tree, she looked visibly depressed. In fact, I'd use a stronger word; I'd say she looked disconsolate. She was grieving the way a human mother would grieve, staring at the ground or into space as if nothing would ever relieve her loneliness and pain.

So it was with Winny and Melly and Davida and Astra and all the others in mind that I was looking forward to receiving the coveted reports from Dr. Muin.

At Kumai, I got out of the still-shiny *Garuda* with the smile of someone who feels known and accepted, who is about to be met by open arms. There was the good doctor in a starched khaki uniform, but when I waved to him he gave barely a glance in my direction. "Hello!" I cried then, in a hale, North American way, but Dr. Muin only nodded. When I mentioned the papers, he said blandly that he'd left them in Tanjung Harapan, very sorry, and I knew, without further signals, that he was upset about something I'd said or done. My mind raced. My last day, my only chance at the reports . . . Whatever it was must have been serious. But what was it? My "instincts" said mea culpa, even as my brain was looking for anything else to blame. Perhaps there were too many witnesses present. Perhaps Dr. Muin had been told not to give me any official information. It certainly wasn't possible that the only copies of those reports were in a bag left casually in a room upriver.

For a few minutes, I just stood there.

Then I remembered our discussion about the rangers, during which I had offered an opinion that disagreed with his.

Leave-taking was strained. A secretary fiddled with her typewriter, and Riska asked me to give her some money to make the typing go faster. We waited for the appropriate stamps, said good-bye to our PHPA hosts, pushed and kicked our packs out to the street, and found a car to take us to Pangkalan Bun and the airport. In the waiting room of the airport there is a curtained-off space with a prayer rug. It stood empty as the hours crept by and the plane did not appear. Maybe I shouldn't have told Dr. Muin the truth about the rangers when he asked me what I thought, but he had asked! I checked with Riska: could that be it? She shrugged. Her thoughts were already on the future, when she would take up her daily job again in the blue polyester dress. "How do you write a book?" she asked. "I keep a notebook of things I remember. From the village. About my people. I would like to write down the real story, not just headhunting but all the culture."

"What about your own story?" It struck me that I still knew very little about this woman with whom I'd lived for ten days.

"No, that's not interesting. I want to tell about the way it was to live in the jungle. The way . . ." She pulled a slim notebook out of her pack and showed me some notes in English.

"Why English?" I asked. "I mean, it's not your first language. To write, you need to get as close to your thoughts as you can. You use the word *jungle* in English, for example. But I usually say *forest*. They mean different things."

"*Forest* and *jungle*. Two words," she said sensibly. "*Utan* and *rimba*. But how do you start?" She had followed me outside.

The little airplane was at last taxiing up to the bit of grass where we stood, its twin engines drowning out our words. Esta and Kristin were lining up on the tarmac and someone had finally gone into the prayer room; there was a pair of shoes at the door. Suddenly, I wanted very much to know that I would be back, flying into this airport, coming off this very plane. I wanted the certainty of that, but I wasn't sure why. The story wasn't finished, whatever it was. "Just write the truth," I said, staring at the curtain pulled across the prayer room door and thinking how little I understood of things under the skin. Life, all of us, everything seemed too fragile. A little truth is as dangerous as a flood. It might propel Riska into a future she couldn't imagine, all knots untied. Why wish that on her?

We climbed the metal stairs of the plane and turned to look back at the two wooden Dayak on the strip of grass and at Riska, so animate, so alive.

"Are you coming back?" she shouted.

"Yes! The reports. I need them. Sure I am."

We flew to Semarang and waited a long time for a plane to Jakarta. There we found a room — one bed, one mattress on the floor — and argued about what kind of food to eat. As often happens in these cases, the open-air place that we chose satisfied none of us, and when we sat down to our last meal together, a stranger came over to chat. My heart sank. For a month we had been idyllic. We had been. For the first time in years. Jakarta was the place I was going to be amputated again. Like Winny, I was losing one of my arms, for that is how my daughters feel to me. Even so, the next day, in all the hustle of the airport, we kissed and hugged and waved and after Esta had disappeared, Kristin and I got on a different plane.

Another Island

Sumatra. It's the only place other than Borneo where orangutans survive in the wild. And like Borneo, it has a reserve where an effort at rehabilitation has been going on for more than twenty years. Gunung Leuser National Park has a population of six thousand orangutans, the largest in the world; however, they are divided into two quite separate groups by a road that cuts through the park. Because the orangutans don't cross the road, the two groups don't interbreed, which reduces the gene pool on either side. Worse, there is talk of another road that would stretch from Bukit Lawang, where tourists gather, to Kutacane in the middle of this park shaped like a butterfly with one torn wing.

Bukit Lawang was exactly where we landed after a seven-hour drive from Medan. Kristin and I had spent the night in a four-dollar room next to a ravine in an old Dutch house that had lost its charm in the process of housing too many Europeans with backpacks. Its barren walls concealed rumpled beds and its bathrooms were slimy and sinister. In the morning, several surfboards, covered in traveling cases, stood in the hall. We stepped over them.

Outside, a three-wheeled cab was being repaired, and we tried talking to its driver after being told that there were not enough eggs for our breakfast. Two or three surfers sat at a table near the driveway eating. Hungrily, we turned to the driver. Would he take us to town? I'd made arrangements for the trip to the national park through a travel agent, or at least I thought I had, but our driver couldn't find the agency. Instead, we sputtered ineffectually up and down the streets of Medan, a town that struck me as exactly the sort of place I would choose if I ever found myself needing to live in Sumatra. We had offered a ridiculously small amount of money for the ride, and since the search for the travel agency was fruitless, we admitted defeat and asked to be taken to a restaurant. Had we persisted, I've no doubt the driver would have carried us and our load of optimism on and on eternally, but we were too hungry to care about national parks and orangutans, so he stopped at a noodle restaurant that provided one woman at the counter who took our order and one in the kitchen who cooked it. Across the street another noodle house was brimming with men.

We had six more days of travel ahead, and I knew exactly how much money was left, but after the noodles, I hailed the first air-conditioned taxi I saw (recognizable by its rolled-up windows) and asked the driver if he would take us all the way to the national park. Blinking and looking somewhat surprised, he agreed and took us back to the four-dollar room for our packs, where we said good-bye to the surfers and drove away again with our chins in the air very grandly. But within moments, we stopped; the driver got out and vanished.

He had gone into one of several small houses leaning together on an otherwise vacant street. A child stood in an open doorway. Three hens and a rooster moved around at her feet. A door opened briefly, then shut decisively. I sat in the backseat of a black car with my daughter. The sun was bright and hot, and aside from the chickens, nothing moved for half an hour. I wondered if the driver had become afraid. Or sick? It occurred to me that he might not know where the national park was and would not like to admit it. Perhaps he was inside looking at maps or perhaps he was consoling a weeping wife. Kristin showed no impatience but looked at everything around us with great interest, whereas I felt helpless and frantic.

After awhile another man came out of the small house and got into the car and drove us away. He spoke no English, but he drove very ruthlessly and allowed Kristin to play one of her cassettes, only complaining when he was stopped by the police and given a speeding ticket for 20,000 rupiahs. There would certainly be no profit in his drive after that, and I measured, not for the first time, the gap between what we could do and what we *would* do, a gap as visible as our bellypacks and cameras. It struck me that as tourists, we are often put in the position of parents when we travel — parents whose indulgence could so easily fulfill a child's desires. But our hosts are not children, and the gap between condescension and assistance may be exactly the one between what we can and should do.

The driver was moody for the rest of the day and showed not the slightest interest in us or the surrounding countryside. There was no pleasant outpouring of enthusiasm about the weather, his country, our country, or anything else. Of course we were limited by language, which has given us great access to other humans except when we don't hold it in common. Then its lack distances us. And we have forgotten all the other ways to communicate, except, perhaps, through music. While Kristin's tape played, I leaned against her and took a long nap, which meant that I was disoriented when we finally stopped. Apparently we had been traveling for some time in a landscape of dust. We left the cab, put on our packs, and walked through a group of tawdry stalls and over a frail suspension bridge. We had given the driver enough rupiahs to cover half of the ticket in addition to 50,000 for our ride. We had seen no one but our silent chauffeur for hours, and suddenly there were people around us, above us and below. Men in sarongs were pulling rocks from the middle of a river and fastening them, in wire nets, to the slippery shore. Children were bobbing in inner tubes upstream of them — one pale child, and the rest dark.

The workers and children in the river were all male, but there were women in the stalls that lined the banks. The women were feeding and watering hundreds of tourists. Or selling trinkets. The stalls were crammed with batiks, shorts and dresses, dusty postcards, bad paintings, slick carvings. All of this seemed to have come from somewhere else. Some factory in Java. There was not a single local invention except for the trash revolving in the river, where children played close to a metal gate across an outtake valve and where two of them had managed to squeeze between the bars. On a rented inner tube one can float from here to someplace else or from someplace else to here, but the point is that this is a place where a tourist can see orangutans, although the seeing of them entails a forty-five-minute hike. A narrow path through the village follows the course of the river and leads to the spot where, twice a day — at feeding times — a leaky dugout pulled by two men takes tourists across, six or seven at a time.

The Sumatran subspecies of orangutan (Pongo pygmaeus abelii) occurs only in the northern part of the island, where most of the forest has, for most of historical time, been uninhabited by human beings. Information about the orangutans who live there is consequently sparse, most of it having been collected by the Dutch zoologist Herman Rijksen, in the early 1970s, when he was formulating the theories and techniques of rehabilitation that have put Biruté out of business.

There were places on Borneo that Kristin and I might have gone during those last six days to see ex-captive orangutans. Wanariset, the site managed by Dr. Willie Smits, was not an option since it is not accessible to tourists, and tourists, as Charlotte Grimm had pointed out, were exactly what we were.

I wanted to stay within the boundaries of Indonesia, where I could see variations of the same policy, variations on a rehabilitation theme. By going to Sumatra, I could also see the visible differences in the two subspecies of orangutan: Bornean and Sumatran.

Most of the tourists in this park were young Europeans wearing sandals or hiking boots and carrying cameras, and, after our hike and precarious river crossing, there were a hundred of them gathered at the other side of the river, waiting for us to join them in an assault on the rugged hillside. Together with two rangers we began the ascent, the astonishing heat already snapping at our faces and licking at our legs. Listening to other tongues mold the sounds of surprise in a dozen languages as the first orangutan appeared in a tree just out of reach, some of us stopped to get a breath and a better look, clogging the path. Our rangers carried a bucket of milk and the usual bananas, and at the end of the climb they crossed a split-rail fence that the rest of us had to respect.

Beyond the fence there is a wooden platform that can be reached by a ladder, in the case of a human being, or by swinging down from tree branches, in the case of an orangutan. As the men climb the ladder, orangutans begin to descend, and they are surprisingly different from our old Kalimantan friends. Their faces have more shape, somehow; they are more delicate and yet less interesting. They seem less like us, I guess, and more like pretty animals. They are on one side of the fence and we are on the other.

Sumatran males do not develop the pronounced cheek pads that males of the Bornean subspecies have. Their faces are diamond-shaped, as opposed to square, and the laryngeal sac is smaller. While males are still twice as large as females, Sumatran orangutans of both genders are smaller than their relatives on Borneo. There has been considerable argument among specialists about how distinct the orangutans on Borneo and Sumatra are. While the difference is genetic, they can mate and produce offspring and have done so often enough. But only in zoos, as far as we know, and now that's no longer allowed. (To complicate things, there may be a third subspecies, in the Crocker Range of Sabah, but zoos haven't begun to worry about this, even though morphological variation is greater between Sabah and Sarawak orangutans on Borneo than between the orangutans of Borneo and Sumatra.)

All 103 of us pushed each other against the fence in order to aim cameras at 6 or 7 of them, subspecies or not. There was some drama on the platform, but it was at such a height and distance that, even with a telephoto lens, it was impossible to record. Because we had no individual's history, we didn't know who was stealing bananas from whom. What we knew was that we were staring at our relatives, and that we had a sudden yearning to take to the trees. And we knew that around this park there is an area of forest that is under attack by loggers, although it provides the only secure habitat in Sumatra for orangutans.

In a report prepared in 1993, the Indonesian government recognized four factors that negatively affect the survival of the orangutans: habitat destruction, fragmentation, degradation, and disease. The report calls degradation "more temporary in its effects" than the others, and admits that it's caused by logging. But wait a minute! Temporary? Herein lies a legitimate disagreement between Biruté, with her focus on location, and the gentlemen in Jakarta, who are in the business of selling timber.

Gunung Leuser was established the same year as Camp Leakey, in 1971. In the early days, as far as rehabilitation of orangutans was concerned, the focus of these centers was simply to ensure the enforcement of the Animal Protection Ordinance and to reintroduce ex-captives to the wild. But the Meratus Rainforest Project report written by the Institute for Forestry and Nature Research and AIDEnvironment in the Netherlands states with absolute assurance that "the original design of the programmes had very serious flaws. Indeed the original design of rehabilitation in a way *added to the risk of extinction* [italics added] rather than served the conservation of the species."

I was not happy in Bukit Lawang, but I could not put my mind quite on the reason. Claustrophobic, it was — during feeding times, the place was clogged like a damaged artery with human beings — and the orangutans were remote. But aren't they supposed to be? Isn't that what nature is all about? Before our climb to the feeding station, we had negotiated a hike that was to take place the next day. We had made a down payment with a guide. The area around Bukit Lawang is mountainous — very different from Tanjung Puting — and I was eager to see it. But that evening, while we were eating, Kristin and I looked at each other and knew suddenly that we longed to escape. "Let's get out of here," I whispered. "I can't bear the thought of another day in this place."

We got on a bus — any bus — and left like refugees, passing a Worldwide Fund for Nature van with a new deportee from captive life, a sad, caged orangutan. Not so long ago Darwin's champion, Thomas Huxley, decided that the evolution of human consciousness allowed nature to become conscious of itself. What does that mean? It means that in primates self-awareness and awareness of others are inextricably linked. Just like Chantek, we assume that other people, animals, and even things have desires, intentions, and beliefs. We *empathize.* I looked around at our companions. The bus rocketed along noisily, covering the hours. I was thinking things like: From the Third World, what cost to enter the First? The gold mine. The forest. Original place. Hajji Makmur in his filthy shirt. Nature and culture. Kristin looks like me at twenty-six. All of this is so right, but not right. "Where you from, Mrs.?" "How are you tomorrow?" "Toronto. America." "Good for you, good for me." I was coming down with something, that was it; I was feverish.

In Tanjung Puting I had dreamed that I was cutting a passage between my house and my neighbor's, which is attached to mine. I dreamed the dimensions of the tiny door I would unlatch and all the rooms I had never seen, my house divided, then joined. If the most disturbing sight on the Sekonyer River was the blanched hot sand of Jemantan, there was another image I had not been able to erase from my mind, that of the three young female orangutans clutching at each other so desperately that they became one creature with six arms and six legs. Divided yet joined. As the bus rocked along, I remembered the photographs in my ex-husband's house of Esta and Kristin and a young, long-haired woman who kept hold of them. In one of them we are naked. In another we have ridiculous, cockeyed smiles. We're staring into the sun, three heads with long, straight hair, squinting eyes, curled mouths. I was young when I had these children. Then my marriage fell apart and we spent the next decade tumbling around, very tight, clutching and clinging. In the face of trauma and disruption and parental disappearance, mutual dependence is the only way to sustain life. The Twisted Sisters are damaged, perhaps even deranged. They may never lead normal lives. Will they mate? Raise children? Can they, should they, be saved?

An emerging branch of biology known as cognitive ethology is proposing that certain species are, in the words of Frans de Waal, "endowed with the capacity for genuine love, sympathy and care." De Waal, who also works with the Yerkes Institute, contends that aiding others "at a cost or risk to oneself is widespread in the animal world." But what does this have to do with saving orangutans?

The big difference between Biruté's rehabilitation program and the one favored by the government is human contact. Biruté thinks, since rehabs are orphans, they should be handled by surrogate mothers. Herman Rijksen, Willie Smits, and the other (mostly male) professionals contend that this makes for an unreliable dependency. At Wanariset, Smits keeps the rehabs in cages, separated by peer group, until they are taken deep into the forest. While caged they are fed and cleaned as mechanically as possible.

Orangutans are shot in palm plantations because they like the tender shoots and hearts of palm. When it's a mother who's shot, the result is an orphan and a huge gap in the natural recurrence of orangutans, since they breed relatively late in life and very infrequently. Like dipterocarps. But orangutans, like humans and unlike trees, must learn from a living, breathing mother. Obviously, our efforts should be in protecting the grown-ups since the young ones take years and years and years to reproduce. I know this. So does Biruté, who talks about habitat (including dipterocarps) but throws her weight behind saving babies.

Then I thought, What would have happened to my children if I had died?

My throat was beginning to hurt. Ahh, this was exactly what I had feared. Far from home, far from comfort or remedy, I was going to be sick and incapacitated. Kristin would have to take care of me. I'd be helpless, like the little frog in the mouth of the hungry snake. In that busload of people, with all our various sins, I saw God manifested as Life: virus, frog, and snake. And maybe it doesn't matter which eats which, whether the snake eats the frog or the frog gets saved, as long as life goes on in all its variety. Variety making life possible because without it, in a one-celled world, life, by definition, cannot proceed. And life is, by definition, procedure. Eating, shitting, reproducing, and being eaten. That is the life to which we ascribe holiness, or at least the life in which we say we believe. As I watched two girls on the other side of the aisle delouse each other's long hair, I thought, as

Homo sapiens, we love not only trees and other creatures, but our own ideas. Culture, the inheritance we pass on, is not only knowledge but specific beliefs. In using or saving. In the privilege of the individual or the primacy of common good. In sacredness.

Does Gistok notice the shape of other beings? Because if he does, he must wonder at the stars, he must delight in the shape of the tree as he climbs it, knowing its springiness, its fragility. And without fragility, where is delight? Where is worship? We don't worship gold or steel, though we fashion images of the sacred from these molten substances. What we worship is the living infant, the great bird, the white ape; we turn the moon and the planets into gods and goddesses with mortal tendencies. While we dream of gold, we worship only life.

Making Advances

This may involve sitting next to each other and casually touching the body of the other.
— Gisela Kaplan and Leslie Rogers

Then it was August in Toronto and I was being called upstairs by Michael to see Biruté on television. She'd been awarded the Order of Canada. "Habitat, habitat, habitat," she said, when Kathleen Petty of CBC asked her what the problems of the orangutans are. "Location, location, location."

"Not poachers?"

Biruté has a way of being immobile. Her eyes, behind the large glasses, seem not to blink. "The destruction of their habitat is their primary problem," she said, her mouth barely moving. "Poaching is secondary. Of course the destruction of the rainforest makes poaching easier. Anyone with two dogs and a blowgun can take an orangutan." When Petty asked whether she was ever afraid, Biruté said gamely, "Personal safety is increasingly a concern for all of us. There are dangers where I work but you could as easily be hit by a bus." This was interesting because Biruté's followers often portray her as a woman in danger, hated by poachers (the forces of darkness) the way Dian Fossey was. The thought of her risking her life for her adopted charges has been part of her charisma.

What about the ferocious bodyguard?

While we'd been traveling in Borneo, the American animal dealer pursued and finally arrested for his involvement with the Bangkok Six — Matthew Block of Miami — had been sentenced to thirteen months in federal prison for the illegal deportation and sale of the orangutans. I learned about the trial when I wrote to the International Primate Protection League, which had been waging a battle against Block through the U.S. court system. Headed by Shirley McGreal, the IPPL is headquartered in South Carolina, and its quarterly newsletter, sent out to members, had made the Bangkok Six and especially Matthew Block a regular feature. A typical headline reads: BLOCK GETS 13-MONTH JAIL TERM — but cell is empty. Over a period of five years, details of the case were reported in many issues, over many pages. "The two-day sentencing hearing in the 'Bangkok Six' orangutan case took place on 15–16 April 1993," ran the story under the "Jail Term" headline. "It ended up with Matthew Block getting a jail term of 13 months, 3 years of supervised release, and a $30,000 fine for what must certainly be 'The Crime of the Century' against animals."

Although there were other parties to the crime, there were no other prosecutions. The trade in endangered animals has been compared with the drug trade, replete with cartels, safe houses, and informers. To me, it seems more like the trade in illegal antiquities because it involves false certificates and the reputation of so-called specialists. Tony Silva, for example, a boyhood associate of Matthew Block's, was known as a conservationist and specialist in tropical birds before he was arrested for selling twelve of the world's last hyacinth macaws.

But, like drug dealers, smugglers are often encouraged to "snitch" on other smugglers or to participate in "stings" in exchange for reduced sentences, which is why Matthew Block "cooperated" with the government on two stings and maybe why Kurt Schafer, a German animal dealer based in Thailand, eventually provided information to Shirley McGreal. He told her that he had been contacted by Matthew Block about a business venture that required the smuggling of wildlife into Bangkok. He said Block knew he had experience moving birds from Singapore to Bangkok and thought he could help untangle the shipment. Schafer flew to Singapore, where the three crates marked LIVE BIRDS were given to him by an unidentified man, and he then put them on his flight to Bangkok, where they were supposed to be transferred on to Moscow via Belgrade. But in Bangkok, the ground crew had become suspicious. Unaware that the crates had been seized, Schafer boarded the plane and flew on to Belgrade.

After their discovery, the orangutans were taken to a rehabilitation center operated by the Worldwide Fund of Thailand. Two siamangs, who had also been part of the shipment, were taken to a zoo in Bangkok. A week later, another illegal shipment was intercepted by Thai Airways officials at the same airport, and this time it was several chimpanzees hidden in crates marked dogs. The flight had originated in Dubai and was scheduled to continue to Belgrade, but the chimpanzees found themselves in a Bangkok zoo with two siamangs, no doubt wondering who, in fact, was in charge of their travel arrangements. In the aftermath of the international turmoil that followed, it was indeed hard not to wonder who was in charge or to comprehend how much the humans involved had to gain by the illegal sale of chimpanzees and orangutans. "Our sociability," Biruté asserted in her TV interview, "is motivated by *greed*."

The plight of the six orphans might have become a drama of heroics. As if there were still two sides to the stage and we were watching *The Ramayana*, it might have been possible to root for the girls and despise the boys who made deals with each other while they quibbled about the value of orangutans. But the actual story was a good deal murkier.

In October 1995, when Matthew Block went off to the Jessup Correctional Institute in Georgia to serve his time, I went off to Kansas to visit my mother, who spends some of her favorite hours at the Topeka Zoo and who had gone to see the orangutans there while we were visiting their relatives in Kalimantan. "You might want to go see ours," she suggested, and I'm afraid I groaned in the way of the condescending child.

"The new man is very good," Mother assured me. So I made an appointment. Yes, why not? In a matter of hours and a few blocks away from my mother's house, I was sitting inside Michael La Rue's tiny office, where he imparted the surprising fact that there are only seventy-seven Bornean orangutans in North America. "And seven of them are right here in your hometown." Leaning across his desk he went on feverishly, as if a clock ticking out the minutes to species extinction was about to strike just behind him. "What zoos are doing these days is species-survival programs," he said. "We cooperate through the American Association of Zoological Parks and Aquariums so that we know who has what and where the best place for an animal is. As far as zoo orangutans go, in a few years ours and the other Borneo ones may all be moved to Europe." He took a deep breath.

When I registered shock, he explained that seventy-seven orangutans simply aren't enough to maintain a self-sustaining genetic resource. "There are a few more over there. In the U.S. we've got mostly Sumatran. So the Sumatran ones over there will come here."

"No one's interbreeding them?"

"Definitely not. In fact we have a 'no breeding' policy for all hybrids. They'll be genetically phased out."

"Because they're too different?" I was thinking that in view of the catastrophe of extinction, perhaps a little mixing was better than no orangutans anywhere.

"They've been separate for ten thousand years. I think there is every reason to suspect they are now different . . ." He paused. "Certainly they have less in common than bonobos and chimpanzees."

"Different . . . were you going to say breeds? Could they be two different races of the same species?"

"That's touchy. I wouldn't want to go into that."

The reason for the "genetic resource" apparently is that in the event a species entirely disappears from the wild, some members of it will exist in captivity. "So that they can be reintroduced?" I asked. "That doesn't seem to work terribly well with orangutans."

"No. That's true. But this is what can happen. Suddenly something you have in captivity that seemed to have not much value becomes endangered. Suddenly they have enormous importance. But we need more research into reintroduction. That's another part of our job."

Needlessly, I said, "We have to find a way to train them first. What to eat. How to swing from trees. The cable they've just put up in the Washington zoo doesn't look like it would fill the bill at all. It's straight and inflexible." I had settled back in the office chair, feeling strangely at home there. Two Kansans in Kansas, talking orangutans. I should have told Michael La Rue what Matthew Block's lawyer wrote to the trial judge after the sentencing hearings: "If ever you actually retire from the bench, thus eliminating the artificial barriers that stand between judge and lawyers, I hope you will grant me a social visit, either here in the beautiful Napa Valley or in Miami. For that matter, I would go to Jerkwater, Kansas, for a beer or two."

"We haven't found a way to build flexibility into zoo structures," my new friend went on. "Orangutans have to learn how to test the weight-bearing ability of branches and all that. And as far as what they eat is concerned, we know so little ourselves. One of the projects I'm most excited about is a research study in Sabah being conducted by Isabelle Lackman-Ancrenaz and Marc Ancrenaz, who are French. They're trying to see how orangutans do in second-growth forest. I'm very interested in this, since it may eventually be all they have."

On my way out of the office, I picked up a brochure called *The New Ark*, which explains that zoos and aquariums are the last refuge many animals have against extinction. "Although their story of saving the world's animal species may seem similar to the Biblical character's, the zoological community's reasons for taking on this task would astound Noah. Noah worried about a flood. Many kinds of catastrophes face today's wildlife. They include human overpopulation (world population increases daily by nearly a quarter of a million people), deforestation (tropical rainforests decrease by one hundred acres per minute) and air pollution (the burning of fuels and forests may cause the atmosphere's temperature to rise three to eight degrees Fahrenheit by the middle of the next century)."

I remembered hearing Biruté say that orangutans would survive in theme parks and nowhere else. I remembered the IPPL newsletter saying: "During his short life (he was born on December 7, 1961) Block has taken the freedom from tens of thousands of animals and put them behind bars. Now he is behind bars himself. Perhaps he will get a taste of what his animal victims suffered." Then I found the path to the Discovering Apes building, following the strong odor that issued from it. Here I learned that the first actual captive orangutan in Europe was a gift to Holland's William V, Prince of Orange, in 1776. They never seemed to live more than a few weeks in captivity, although Lord Monboddo, the eighteenth-century Scottish philosopher, declared that orangutans were qualified to be described as human beings and would, in time, learn to speak.

The Topeka orangutans did not look as if they would ever learn to speak. And they did not remind me of my friends in Tanjung Puting. For a moment I wondered if they weren't Sumatran after all. Could a zoo be mistaken? There are blood tests and so forth, but how could anyone be sure about these things? What interested me about the zoo orangutans was that the young ones weren't anywhere close to the mature female who sat on a concrete branch staring morosely through the glass. She'd been described as the adoptive mother of one of them, and yet I did not see any responsiveness at all between them. Two of the youngsters were chasing each other around and tangling themselves up in a rope that hung from the concrete trees. The exhibit was entirely glassed in, and while I stood there a flock of schoolchildren came in, laughed, shouted, tapped on the glass, were appropriately thrilled and horrified, and then went out again.

None of the orangutans they'd watched had been born in the wild. There was that much to say for the zoo's efforts at conservation, at least, although the thought of studying second-generation slaves in order to understand African villagers occurred to me. But zoos always make me sad. I probably looked as morose as the female orangutan on the other side of the glass. And there's no denying that some zoos are stocked with animals taken out of the wild illegally. Matthew Block made his peace with the U.S. government by taking part in a sting designed to catch five Mexicans (including one zoo director) who were lured to Miami to buy a baby gorilla for a Mexican zoo. The sting involved a shipping crate loaded on a plane with a wildlife agent dressed in a gorilla suit inside the crate.

On my way out, Michael La Rue had told me with some satisfaction that Biruté was coming to Topeka to give a lecture. "She has friends here," he said, "though not the scientists, of course."

Since I was going back to Toronto, I begged Mother to go to the lecture, but she refused very pointedly. "I don't like her. She's not nice to you," she explained, which reminded me of Biruté's mother's protectiveness. "Why don't you leave my daughter alone!" she'd snapped, the last time I'd called the OFI. But a week later my mother told me enthusiastically, by phone, that she had relented and gone to the lecture and that Biruté had been "charming. All in black and very neat, although I didn't think the slides were very good. But maybe it was just my eyes. She handles herself well. She made a joke about not being as technically proficient as an orangutan when she had trouble with the mike."

"People liked her?"

"Oh my, yes. She had a very good crowd and she spoke for an hour and a half. You've heard her, you already know all of it, but I was really upset about the man who's doing all the logging."

"Who's that?"

"Oh, I didn't write down his name, but he's a big billionaire who has the concession to cut all the trees and he gives a lot of money to Prince Charles for the homeless and to Worldwide

Fund, so nobody will touch him. She said the government's fine, but there are some individuals with too much greed."

"What else?"

"She showed pictures of her husband and said how good-looking he was when she fell in love with him and that he still is. It was quite sweet."

"Did she say she's no longer working with orangutans?"

"Oh, not at all. No. She collected lots of money for them."

Matthew Block was serving his sentence in a Georgia jail, and Biruté was raising money in Kansas. There were, perhaps, thirty thousand orangutans left in the world.

A View from the Trees

I had already decided to go back to Borneo, but now I decided something else. Even though I hadn't heard a word from Biruté since the evening of our dinner in L.A., I wanted to lay eyes on her once more, and if possible, I wanted the laying on of eyes to take place close to the forest from which she'd been exiled. I would go back when I was sure she would be in Kalimantan.

Riska and I exchanged letters, I heard from Pak Herry, head of the park, and also from Dr. Muin. In fact, whenever the phone rang in my office at 3:00 A.M., I knew it was someone in Kalimantan sending a fax. In this way I learned that Astra — the little orangutan we had first seen in a box with a rag, the one we'd fed with a bottle and who had made Esta sick — was getting fatter and had been moved to Tanjung Harapan. He was in a cage at the brand-new clinic, getting used to the new surroundings, and there were two new orangutans in there with him. "People gave those orangutans straight to the office," Dr. Muin wrote, "because they know that they are protected. After getting some health examinations and vaccinations we will set them free in the forest jungle within the park because they are indeed wild animals."

The exchange of information was exhilarating. I'd begun to notice that most of my friends were giving up one thing after another and taking up one thing after another: meat, fats, wine, cigarettes; exercise, antioxidants, melatonin. A city is like the prairie, where everything is hard-won. None of the ease of the tropics here, but bitter edges and every tree an individual to be cherished. I had come home in July. By October, I wrote back and said I was returning. I was beginning to yearn for the forest again. Were there any requests? What should I bring?

"Bic lighters for the rangers," Riska wrote. "And a pair of Tevas for me." No requests from Pak Herry or Dr. Muin.

One day, in the yard across the alley, two men were trimming a tree in my neighbor's yard. It was a spruce that stayed bushy and green all winter, sheltering the birds who come to my feeder and then rest on its branches to congratulate themselves and communicate with each other. I had been talking to Anne Russon on the phone — she was just back from Wanariset — but I put the receiver down and ran outside to investigate. "Trimming it?" I asked the men.

"Taking it out. They don't want it anymore."

In addition to its luxuriant beauty, the tree had protected my neighbors from a view of my kitchen, and me from a view of theirs. More rooted to the neighborhood than I or the people who lived in the other house, it had no way to defend itself.

Now, its loss seemed as unbearable as the loss of a relative, although I had taken it for granted until that day.

"I'm really sorry," I said to Anne, when I picked up the phone, "but there's something I have to do. Come by with your pictures. I'll talk to you then." But instead of tying myself to the trunk of the tree, I got in my car and drove away, knowing that when I returned, the tree would be in pieces in the back of a truck. Saving the tree simply involved too much intensity, too much interruption in my own life. I would have to find my neighbors and become, on some level, intimate enough with them to convince them that the tree had rights. Or that, anyway, I loved it. Maybe I didn't really care about the tree. Maybe I only cared about myself and the view from my kitchen. Or maybe my romantic, woodsy view was obstructing their sunlight. It was way too complicated and probably way too late.

My niece, who was visiting when the tree was cut down, said, "There's a woman in my office who's a real lefty, a big liberal, but she had four trees taken out of her yard. I couldn't believe it." This started me thinking about our expectations. I, who'd already begun to hate my neighbor — and *hate* is a word I stumble over — could forgive Hajji Makmur, who poured mercury into the Sekonyer River. Was it because the tree was mine in a way that the river isn't? Of course not. We are tied to the natural world, to forests, and rivers and trees, as much as we are tied to each other. And we're hurt by the sight of a fallen thing. Because we understand death, we know that a tree disappears forever, that something is gone that will never come back. A hole is a terrible thing. A polluted river still flows; we can fool ourselves into thinking it's healthy, but an orangutan or a tree is different.

When Anne came by to show me her photographs of the Wanariset orangutans she had been studying in East Kalimantan, I realized that our relations with nonhumans are as subjective as everything else in our lives. We see them according to our own experience. As I walk Esta's wild-looking male dog, the reactions from people on the street are astonishing. Some of them soften immediately as they see his cock-eyed stance and tilted head. (He has neurological problems.) Others give him a wide berth or mutter disapprovingly. One young woman leaped into the street and screamed, "Oh Jesus, help me!"

Anne's orangutans are as distinct as a group of *Ramayana* players, or maybe I see them according to the light she casts them in. Unlike the orangutans at Tanjung Puting, who, up to the last day, remained indistinguishable as far as I was concerned, these images of orangutans had very distinct characteristics because I was seeing them through Anne's lens. One reminded me of my father, with his stern eyes, receding hairline, and wide forehead. "He's called Charlie, and he was the terror of the forest," Anne said. "Do you see this little female. This is Imelda." She pointed to a pathetic-looking creature hunched on a fragile branch a few inches from where Charlie sat glowering. "He held her captive for weeks, keeping his eyes and hands on her so that she dared not move away. She was repeatedly raped. She looks cowed and terrified. The males are afraid of Charlie, too, although one came to fight for Imelda, and Charlie fought him off and then rushed after him to chase him away. After they'd been gone a few minutes," Anne confided, "Imelda suddenly sat up straight, made a gleeful smile and dashed away."

"Any photos of DiDi?" I asked, because ever since I'd watched a TV special about Willie Smits months before, which featured a Taiwan family–raised orangutan, I'd been interested in DiDi, who had lived with the family for seven years. They dressed her, took her out in a stroller, gave her noodles and meat and her own bed. She played the piano. She went to restaurants. When her humans gave her to Willie Smits, the parting was unbearable to watch. While DiDi was in quarantine, Smits discovered she tested positive for tuberculosis, and the resulting seclusion set her progress back. DiDi is an animal who might become what Anne calls bicultural, except that she has to learn how to be an orangutan — to know what trees and branches are safe, what leaves and fruits and nuts to eat, even how to climb and swing. And a cage is no place to learn this kind of thing. But one day, after months in solitary confinement, DiDi was at last taken on the long walk into the forest, a turning point so important in her life that her human Taiwanese mother was invited to fly to Indonesia to witness it. With the magic of television, we could watch both of them, and we saw DiDi reach out for her first real honest-to-goodness branch, grab it, and disappear. Then we heard a crash. Her reach into the ancestral forest had been a disaster. She might have been killed. Instead she was released into the wild, where Charlie was still asserting himself. And like Imelda, DiDi was held captive for weeks, during which time she was not allowed to eat. Or rather, she only knew how to eat bananas, and since Charlie took all of the ones the rangers left in his area, DiDi got hungry and dangerously thin. Anne stared at malevolent Charlie. "Did you know he was the only one of the Taiwan Ten not orang-napped by Biruté's Indonesian students?" she asked. "And that's because he was really rough even then, and they couldn't get him out of his cage. They did try, but he wouldn't go along with them."

I knew only a little about the Taiwan Ten and how, soon after the Bangkok Six, they had landed Biruté in the headlines a second time. Owning an orangutan is legal in Taiwan as long as it's registered, but once grown the "pets" become dangerous and destructive and are often abandoned or kept hidden away in a cage. These ten had been chosen for repatriation to Kalimantan and "progression to the wild" out of three hundred or so who were registered. Schoolchildren had sung farewell songs as they were loaded onto a plane. News photographers had taken pictures. They were bound for Wanariset, the new station in East Kalimantan managed by forester-turned-conservationist Willie Smits.

Anne said, "It happened while I was away, but Biruté claimed it was her people that set the whole thing up, the repatriation from Taiwan. I'm pretty sure she was in Taiwan herself orchestrating the transfer, and since many Indonesians don't speak English, they were probably just as glad to have her do the legwork. Anyway, the ten orangutans came to Jakarta and were housed somewhere. And probably because of the terminally slow process of Indonesian paperwork, they didn't get moved immediately and that gave Biruté a chance to grab them."

Some of Biruté's students at the National University of Jakarta had intercepted the ten orangutans on their way to the forests of Wanariset and taken charge of them.

"I think it was after the Taiwan Ten that Biruté seriously lost touch with reality," Anne said. "They took those orangutans hostage and held them in a house in Jakarta for a year. I mean, her students did it at her direction, and it was very dangerous for them. I went down to the Ragunan house several times when I was in Jakarta — a couple of times with Mark Starowicz from the CBC, when he came over to make *The Third Angel*. In fact, I've got pictures of Mark and his kids with the Taiwan Ten! The incident when Charlie got left behind happened when things really heated up, when the government people barged in to test the orangutans for TB, and there were open disputes about where the ten should go."

Eleven months and several (human) arrests later, the orangutans were finally confiscated by the police and moved to the destination originally intended for them. Or seven of them were, including Charlie. Two of them had contracted tuberculosis. One of them had hepatitis B. Those three had to be sent to the Primate Institute in Bogor on Java.

"The PHPA claimed it was Birutė's care that was at fault in all these infections," Anne said. "But she claimed the people who came to 'test' them had infected them deliberately in order to have a reason to confiscate them!"

"And what do you think?" For the first time I could sit in my living room and talk to someone about animals we both knew half a world away. I felt that I had joined the ranks of the involved, which is probably the feeling all ecotourists come home with. At any rate, I could understand most of what Anne was saying.

"That Rangunan house had very little security and a lot of students came and went quite freely. Visitors came to see the orangutans and often handled them and there seemed to be little control over the state of the food or water they were fed. I mean, the kitchen and food storage areas were quite dirty when I saw them."

"Couldn't the government have stepped in and just taken all of them earlier?"

"Sure. Except that Indonesians are always very *malu* about conflict-ridden behavior and just plain aren't brave enough to confiscate if someone says no. Physically it wouldn't have been hard. They were all relatively little guys, max about six years, and other than Charlie and maybe one other tough guy, you could simply have picked them up and carried them away in your arms. Their cages were also portable, so it wouldn't have been very difficult to load them on a truck. But socially and psychologically it would probably have been very hard. Finally the students decided to bring things to a head, and the story I heard was that they loaded all of them (except Charlie, who declined the invitation) into taxis and hid in the streets of Jakarta. I think it was two or three days before they decided to give themselves up. So Charlie was with the others except for this taxi chase . . ."

"But he got sent to Wanariset, too," which was what Biruté didn't want. "Was Wanariset the best place for them?"

"Willie's rehab program was the only alternative. Remember the new edict. With Willie, rehabilitants are sent into the forest with a good layer of fat on them and with a group of peers, but that's about it. Most of them, according to his records, survive. But release records don't always agree." Anne closed the book of photographs. "After the Charlie episode, DiDi was recaptured and put back in quarantine," she said, looking down at it. "With an infected wound. And she is not angry and upset at being behind bars anymore, which is the saddest part of her story."

The new edict was that statement I had read, framed in its ominous black box. The second paragraph of it said:

> . . . The rehabilitation process has become a means in itself, generating cash, yet continuously endangering the wild populations in the surrounding through their invitation of disease transfer from visitors to wild apes and vice versa. Because a new strategic approach is available, the outdated rehabilitation stations can presently be closed as instruments facilitating law-enforcement; indeed the Ministry of Forestry has decided to terminate the former way of rehabilitation and proceed according to the new approach. This implies that the old stations at Bohorok and Tanjung Puting/Tanjung Haragan will be closed for new entries. The quarantine facilities at Wanariset, East Kalimantan, will be the exclusive quarantine and headquarter station of the orangutan survival programme for Kalimantan.

Wanariset is run according to the new rehabilitation credo, and its principles are as far from Biruté's as it is possible to go. Most startling, perhaps, is the rule that no tourists are allowed. Over the next few weeks, I was to learn more about the differences from Anne, who has been working there, off and on, since she left Camp Leakey. She told me that orangutans there are kept in a quarantine facility for at least two weeks and are socialized in peer groups with minimal human contact. Crucial to the program is the fact that they are released nowhere geographically close to wild orangutans. They only have to compete for space with each other, whereas Biruté's orangutans had been released individually, deep in the forest, where they presumably feralize, but where they may force wild orangutans to search for a new range. At Wanariset, Anne said, a group is stationed in a cage at an uninhabited location. After a few days the group is released and provisioned daily for as long as they need provisioning. Subsequent groups are placed in new uninhabited locations at least a kilometer or two away. Diseased, unwilling, or dangerous individuals are taken back to quarantine, the way DiDi had been.

"It's still hard to see anything through," she admitted. "When a vet isn't at Wanariset almost daily, orangutans start to die. Once Chris Warren, the regular vet, left for a couple of weeks on holiday and when she came back she found almost half of the orangutans at the clinic seriously infected with scabies. Part of the problem is that, like at Tanjung Puting, routine daily care is handled by local staff who don't have the medical expertise to handle the problems that crop up. Their education is pretty good by local standards but most have no medical qualifications or experience caring for wildlife when they start work and their cultural perspectives can be at odds with the conservation aims of their jobs. Obviously it's also hard to get local staff to provide the high-quality care the orangutans need, like expensive medicines and carefully balanced diets, when that care is better than what they can afford for themselves."

When Anne studied a group release at Wanariset, she noticed enormous variabilities in success rates. If motivated, the orangutans picked up skills rather quickly, either by working things out on their own through experimentation or by studying the methods their peers were using to get at food, build nests, and navigate the forest. But the ones suffering from what we humans might call "low self-esteem" lost track of their peers and seemed reluctant to set out on their own. I remembered what Lyn Miles had said about motivation in the learning of sign language, but I was not ready to frame my questions about any of this. I wasn't sure what the questions were. How do we learn? Is it different for orangutans? If language isn't the thing that separates us, what is? If we can't guarantee self-esteem for human children, how can we guarantee it for orangutans? I looked at Anne's wonderful photographs and read her reports and went back to my long-distance letters.

Pangkalan Bun, 11 Augustus 1995

Dear Linda and all the girls

How is the season now in Canada? Is it summer?

It would be nice to know other countries so I can imagine (a little) how nice it is. In our country, now is the dry season but still every day we have lots of rain. Sometimes in a day we have heavy rain and then it is followed by extremely hot weather. Very unpredictable weather lately.

By the way, do you remember Mrs. Charlotte? I met her in the airport one day and I told her a little about you — that you are a friend of Dr. Anne in Canada. And when I explained a little thing to her, she said that she was very very sorry and uneasy because she behaved so rudely to all of us. She said that she was just very emotional after she heard somebody's report and just rushed to us without thinking twice. Now Charlotte looks very friendly to me (I do not know why). This makes me a little surprised because (mostly) she kept her distance from us before. But everything is fine. How is

Kristin now? Is she back in America already? And how is Esta? I hope everybody is always happy. Everything is just like yesterday when I remember how nice it was to stay with you together.

Easy going and nice picnic.

By the way, enclosed please find some photographs about the Dayak area, long house and the Dayak Tumon dancers. I hope you will be pleased to see them. How is your planning to come in January? I hope you can manage to come soon. It is still very nice here and the baby Astra is still very sweet and heart breaking. Please make sure to see him again.

Love, Riska

I made a list of things to buy and pack. Another pair of river pants. Sneakers. Poncho or rain jacket. Hat. This time my trip would be during the rainy season. This time I'd do everything right. Bigger notebook. Faster film. We may be the last generation to travel between one age and another, since it's possible to take a plane from a twentieth-century city and land only hours away from the Stone Age. There we can meet people who have never seen a plane, except in the sky, and who have never seen the ocean or a bulldozer or a high-rise. We become emissaries of the future and our responsibilities are much heavier than the responsibilities of an ordinary diplomat. We are not trying to sell the virtues of our regions but to warn against the ruin we have brought to them. We carry Bic lighters and messages of caution, but our machine-made clothes, our health, and the apparent control we exercise over our fates give us away. Instead of being eyed warily, we are envied.

To make matters worse, we're confused with our compatriots who have preceded us and who are even now making a killing out of the few unspoiled places that are left to be spoiled. They're making their nests, laying their eggs, growing rich and fat, surrounding themselves with walls and servants and bodyguards. They have power. So how can we preach? What about flying out of the urban twentieth century into a rainforest full of orangutans, proboscis monkeys, frogs, and snakes? Do we see this earth as a piece of work to be guarded and mended or only visited?

Riska was promising me a Dayak village and that, truth to tell, was the biggest incentive of the second trip. The Dayak have lived in the forest for thousands of years and they share it with thousands of other species, among them orangutans. Maybe they still know how to live as part of nature, as part of the garden, as participants. If so, I wanted to see how they do it. It was part of the follow, as I saw it.

One night, walking the dogs, I stood outside one of the houses in our neighborhood. The curtains were open at the upstairs window so that I could see a room with dark moss walls and white ceiling. One of the windows was cranked open and a crib sat right in front of it. If I could have the mossy room and the confidence it announces, I thought, I might not be a wanderer. I can't figure out what pushes me away from all certainty. Is it the crib that gives a house its anchorage? Another house, just where the alley meets the street, is being rebuilt by its solitary owner. It's an enormous job and he's doing it by himself, a Northern individualist. I stared at this house, as well, and thought about the European ancestors and what they have done to us.

Second Sight

When I emerged from customs in Jakarta, sleepy and disoriented, and this time very, very much alone, I stumbled out to the taxi stand carrying a shoulderpack full of gifts, a bellypack, camera bag, and a large, unbearable backpack full of clothes, books, a tape recorder, and a fleece jacket. "It's going to be chilly," Riska had warned. The books were *Uncle Tom's Cabin* and *Emma*, which I thought would keep me occupied during long rainforest days. The gifts were for the PHPA rangers. Carey and Trevor had told me it was one thing I could do, as a long-term guest, although gifts of money were inappropriate, and I had still received no requests until two days before my departure. Then Dr. Muin had actually sent the park reports I wanted, by fax, and he and Pak Herry wrote long letters. Between them they had hatched a wish list that turned my last hours at home into a strange-indeed shopping adventure. Here is the fax I sent back:

> Dear Dr. Muin:
>
> I received the fax this morning and now I'm trying to track down one or two of the things you have suggested, but I need a little

> more specific information. I spoke to several people about the reagents for blood tests. But do you require testing for specific strains of hepatitis? Such as hepatitis A, B, C? If so, this is very, very complicated according to my sources, but I can bring a kit for liver function testing. Would that be adequate? I may be able to find a stain (?) called Gemsa (?) (I think), which is used to determine malaria. Is that what you had in mind? Also the blood pressure cuffs should be possible. I will try to find an adult (human) size and a pediatric one. By Day Pack, do you mean a regular backpack? And what are head batteries? You can see that I'm very ignorant, and I do apologize. Apparently the tranquilizer gun is impossible to acquire and I might end up in prison somewhere if I tried to carry it on a plane. Please forgive me if I don't arrive with this . . .

The biggest item on the wish list was so far from my experience that I did not even know where to begin. Then Esta's beloved Doug had rescued me and I was now bearing, along with everything else, a piece of late-twentieth-century white magic called a Global Positioning System.

Surveying all this paraphernalia, a taxi driver said sympathetically that he couldn't understand why I was going all the way into Jakarta. Why not stay closer to the airport?

I explained that I was flying out of the smaller airport.

"Going to where?"

"Kalimantan."

"Impossible," he said. "All flights to there are from this one."

Since I had a hotel reservation and a shred of self-confidence, we made the long drive into Jakarta anyway, where he followed me to my room, sat down on the bed, and picked up the phone. My daughters chastise me for not erecting boundaries, and there I was with a strange man sitting on my bed in a city where I knew nobody. Worse, I was unable to ask him to leave. Instead, I began to pace, unpack a few things, and use all the body language I could muster to demonstrate my desire for solitude. I longed to sit down, but I would not even take off my shoes for fear of providing what might look like an invitation to a sexual escapade.

At last, the stranger in my room stood up, announcing that he had straightened things out for me. The letters HLN written on my ticket meant Halin — the smaller airport — and he would be back for me at six-thirty the next morning.

With that, he exited.

I got up at five-thirty, called my mother, went downstairs for breakfast, saw my advocate waiting by the glass doors and invited him to have a coffee. It was beginning to rain, so we decided to leave a little early. I ran upstairs to get my things, but by the time I got back to the lobby, the rain was pelting down, and since the hotel is on a rise, people were gathering on the steps outside, looking up at the sky and down at the ground with an air of finality that alarmed me. For half an hour or so, the open sky continued its assault. In the street, two cars floated past and a man in a poncho pushed a motorbike through water that covered the handlebars. I told anyone who was listening that there would be no more planes to Pangkalan Bun for a week, that I would be stuck in Jakarta, that this was a catastrophe, but suddenly my advocate pushed me into his car, threw my pack in the trunk, and we plunged resolutely into the 1996 Jakarta flood. Water came up to the car windows. All this had happened in less than an hour.

Slowly, we plowed through the flooded streets, while men with umbrellas waded out in water over their waists and measured the water level in various places with long sticks. Pointing here or there, they directed us to the highest parts of the road. Nothing was moving now except the heavens and the men with umbrellas and the car that carried us. My new friend's expression had the hint of secret humor I like. He was making this effort on my behalf, even inviting me, when I came back through Jakarta, to stay at his house. "I am alone," he added, as if this would give me confidence.

At the airport, a lot of people sat in the waiting room. It continued to rain. A plane landed. Crew members in crisp blue suits went out to meet it, holding umbrellas. Those of us waiting inside were given small boxes with slivers of cake, some juice, a meat dumpling. Our departure time changed by fifteen-minute increments. In a tiny office with a bucket catching water from the ceiling, I was told that no one could promise me anything. This I believed. Or it might be possible to fly to Ketapang. Where? As a matter of fact, the flight to Ketapang was just boarding.

Ketapang's airport is a slice of the past. There is a small building surrounded by a white picket fence. I'd almost forgotten the sensation of being lost and found at the same time — the walks I used to take with my old, deaf dog that led us away away until we no longer knew where we were. A stewardess wrestled with the door of the plane, and I climbed into a landscape where no one expected me, where I could not use a phone and where it was not raining. In Ketapang, apparently no one spoke a word of English. I was the only tall person, the only outsider, and someone was loading my packs into the back of a pickup truck. Who was he?

"Do you speak English?"

"No. You speak Malay?"

"No. *Español.*"

"Español?"

"Oui."

After a long drive, a wooden hotel, big and square, loomed up at the end of a road, and I got out of the truck with a measure of relief. It was still possible that I was being kidnapped and held for ransom, but I was shown a card with various prices for rooms, and for the equivalent of seven dollars, I was given a key and directed to the stairs.

Within minutes, I'd washed out my undies and T-shirt and hung them in the moist, unquiet air, seen a gecko run for the open window, heard music coming from next door, and looked down on a porch with plastic bottles, a colander, and a dog. I was experiencing bliss. Taking all my clothes off, even the waterproof watch, I threw several pails of cold water all over myself. I didn't want to go out, to break this spell. The fan, the mirror by the bed, the grille over the door . . .

Outside there were painted houses with their tin roofs glowing, with their narrow boardwalks over ditches in which families bathe, with not a child crying anywhere but lots of laughing, as if a tremendous joke was in the air. Sounds of soap operas, pounding of a pestle in a mortar, a cough, a voice. Hundreds of swallows sucked in and out of an empty building. Living here, how would it be, with four or five kids and a chicken and nothing to do but cook and wash? At five, the papas come home to take their boys out for bicycle rides. A gang of teenagers sits by the road, one guitar, one bottle of arak on a stump in an empty lot. "Hello, Mister!" I've forgotten every word I learned except for *selamat* and *nasi goreng*. It's hot. There's a be-still tree in the road.

Tree-of-the-beautiful-name. I want to die in its arms.

Sitting at a table facing the street, no wall between anything and me, I ordered a bowl of something. A man with a cart came in to see the cook, then stopped and held something out to me. Green mangoes. At my smile, he filled a bag. There was spicy salt folded in a piece of newspaper, so I had my desserts and carried the meat from my rice dish home in a napkin to fling out the window of my room. In a moment there were two dogs, then three. It was six o'clock and soon it would be too dark to read *Uncle Tom's Cabin*. Nothing like solitude. Nothing like anonymity. Another green mango please!

It seemed entirely possible that I would never be found, never be released from this dream, but at seven o'clock the next morning, an elderly man arrived on a bicycle, got off, came through the hotel door, and pointed at me. Then he got back on the bicycle and rode away, and a few minutes later a car stopped outside. Fate brought me a forty-five-minute ride in a nine-passenger plane. We skirted the top of the rainforest. Sometimes in the greenest, thickest parts there were thin trails for loggers made of crisscrossed logs that looked, from above, like tracks for a delicate train.

In the distance was the airport in Pangkalan Bun, with its two wooden Dayak and its excessive white floors. Knowing that it was ahead brought a thrill of familiarity, which is quite different from other thrills. That sense of anticipation mixed with possessiveness causes my fur to ruffle when I smell certain places or feel the swerve of certain ground underfoot, but I hadn't expected it here.

The Blue Kecubung had doubled in size and now featured a cascading stairway, artificial flowers at every turn, and an illegal sun bear locked in a cage at the rear. On the first trip, I had gone back to visit him, but a second visit would not, in this case, have brought any thrill of return. Instead, I would have hated myself for failing to come to his rescue in the same way I had failed to save the back-alley blue spruce. On the counter in the lobby, I picked up a glossy brochure.

GREAT ADVENTURES WITH THE GREAT APES OF BORNEO

> See the Endangered Orangutan in the Wild: No less than 750,000 acres of equatorial swamps and tropical rainforests have been conserved and managed by the Indonesian government's Department of Forestry, PHPA branch, and partially funded by the ORANGUTAN FOUNDATION INTERNATIONAL of Los Angeles, under the capable guidance of Dr. Biruté M. F. Galdikas since 1971 . . .
>
> Beautiful photos of the Rimba, a young Biruté

with (orangutan) child, and a close-up of a bright red seafood lunch.

In the basement restaurant, I gulped down two cups of coffee before Riska walked in. Someone had told her about my arrival. I stood up. She sat down. We laughed and felt stunned. I had come back. We were going to spend three weeks at Tanjung Harapan. "We'll both work," I promised. "This time on two books!"

It was Ramadan and the muezzin was exhorting us, through a horrible loudspeaker, about the perils of being with men. "A good Muslim woman," Riska translated, "is . . ." then she interjected, "a victim!"

More laughter.

All night the exhortation went on.

The next morning, in Kumai, I took my offerings to the PHPA, including the little box of magic that allows rangers to position themselves very exactly in the forest by bouncing signals off satellites and triangulating their position. When Dr. Muin and Pak Herry had sent the wish list, I'd asked several people what GPS meant and been told things like: "medicine" . . . "glucose" . . . and "call the Metro Zoo." Once put right by Doug, who even found a source on the Internet, I ordered one that was delivered by courier from Maine an hour before I left for the airport. It was packed among the other gifts: flashlights, sterile gloves, cuffs, kits, seven backpacks, and an English dictionary. Now everything was formally received by Pak Herry's assistant, Mr. Subo, who took me into a private office as if I were a visiting ambassador, but left Riska standing outside. Hierarchies again. Pak Herry was away in Taiwan with Willie Smits, and his replacement was disturbingly good-looking, so I crossed my legs, tried to be quiet, tried to sit up straight, as my mother would have wished. I had managed to find child-size blood-pressure cuffs that were prettily decorated with pictures of Donald Duck. I had failed with the blood-testing kits, although, after much effort, I'd come close.

Only when I asked about Mrs. Biruté did Mr. Subo pull back. "Does she still keep orangutans illegally at her house?" I said, perhaps hoping to goad him into a candid reply now that I'd laid my gifts at his feet. (Besides, I was annoyed at the treatment they'd given Riska). The PHPA complains about Biruté, but they are the only ones who can stop her doing this. Why, I wondered, don't they simply intervene? Unless it's another case of that *malu* Anne had told me about, a cultural training against open conflict. It wasn't even the PHPA, I had learned, that revoked permits; it was another agency altogether, known as LIPI, which issues the permits for scientific projects. Mr. Subo said nothing. Perhaps he did not speak enough English to understand my question or to reply. But who should appear at that moment but Dr. Muin, who seemed as surprised as Riska had been to see me, despite all of our correspondence. Maybe people often come to Tanjung Puting and make a big fuss about the orangutans, then disappear.

"She still does," he assured me, in answer to the Biruté question. "We have confiscated two but she still has others at her place." Dr. Muin threw up his hands, adding that there had been two film crews on the river recently with Biruté, and that he had asked them not to film her with the ex-captive orangutans in the park as if they were hers. "They should be seen with our rangers," he insisted with his uniquely emotional voice, admitting to me that a filmmaker can do the kind of editing necessary to make things appear exactly the way they want them to appear anyway. "And now more films crews are coming. With a movie star. To make a movie of her life!"

"Sharon Stone?"

"Roberts, I think. Something Roberts."

I relaxed. Even uncurled my legs.

There were seventy-five ex-captives in the park now, and in spite of the decision to settle new ones elsewhere, they were still coming. In fact, four more orangutans, all confiscated from one village, were due any day. Yes, they should go somewhere else, but where? Both Dr. Muin and Mr. Subo shook their heads. The present locations are overburdened, they explained. Mr. Subo made a formal speech before I left, and I answered that I was grateful for the hospitality of the PHPA, since they were allowing me to stay in a dorm at the first station. When I stood up, it was because Yadi had appeared in the outer room, sporting a thin moustache. "He's a father now," Riska explained.

In a moment, we were climbing aboard the beautiful *Garuda*. It had been reserved for Biruté, but she had fortunately canceled at the last minute. "She's coming up in two days. With a movie star," Yadi explained to Riska, who translated for me.

"Julia Roberts," I said.

Yadi looked pleased.

Having the *Garuda* meant that we could settle ourselves at Tanjung Harapan, then go up to Camp Leakey and be back at the first station when Dr. Muin came up to deliver the four newly confiscated orangutans.

"It's great to be back."

We were carrying three passengers. One was Mr. Jacki, Riska's friend who managed the Indonesian branch of the Orangutan Foundation in Pangkalan Bun and worked closely with Biruté. The other two passengers were Indonesian students being dropped at an abandoned station downriver from Camp Leakey to do research on proboscis monkeys. On the deck the chattering of palms and pandanus surrounded us, as if they were host to ten thousand cicadas instead of being ten thousand strings against which our motor played. And since coming back to something, as I had discovered, is better than discovering it and discovering it is a wonderful thing, I was aware of my happiness, which isn't always the case. Each picture signpost on the river — the broken flute that means "No Music," the burning water that means "No Fire" — made me a belonger. I knew how to read the signs. I spoke the language. I remembered the bend in the river after the second fire sign and the hut with the sinking dock where we once spent the night. Crocodile Homestay. I couldn't have been happier, unless I'd been allowed to bring back my dead.

Mr. Jacki was going to help the students establish themselves and then return to Kumai with Yadi. I asked him to carry a letter back to Biruté, who had been in Australia but had returned. I had timed my trip perfectly. She would be coming upriver. "With a visitor. Very important. A movie star," said Mr. Jacki.

I wrote a quick letter in my spiral notebook, tore it out, folded it neatly, addressed it to Dr. Biruté Galdikas, and handed it to Mr. Jacki.

Dear Dr. Galdikas,

You may remember me from several letters, phone calls, and a dinner we had in Los Angeles. Well, I took your advice and came to Tanjung Puting last summer with my daughters. Now I'm back, six months later, and I'd love to see you again, as we discussed. I could meet you anywhere on the river (I'll be here for three weeks at Tanjung Harapan) or at your house in Pasir Panjang.

Very sincerely . . .

"Please give it to her," I told Mr. Jacki, in English, while Riska translated. "Please. The minute you see her."

Now the excitement was intensified. In a few days I would see Biruté in context. I would see her with her familiars, the orangutans she had raised. We left our three passengers at the abandoned station trying to refloat a sunken sampan. Drowned boats litter this water; it keeps them from floating away, from being stolen by an escaping orangutan.

But doubling back in our own wake, we came upon the first station, seeing its lime green buildings among the trees and Gistok rolling slowly toward us, and I was unaccountably let down. This was it. A little patch of ground with five cabins. Nothing special. So much preparation had gone into this second arrival, so much life had been lived so far away, and yet Gistok had been there all the time. For Gistok, nothing had changed. After the first flush of recognition it seemed as simple as that. Life goes on in spite of our absences, if not in spite of us. While I had been fighting off the blizzards of Canada, Gistok had been rolling himself up and down the grass in front of the cabins and pulling Alan's clean clothes off the line.

Alan was still the ranger in situ, and he and his coranger had already managed to fill the water pails in the cabin and cover the mattresses with one clean sheet each. But, as he helped Yadi tie the *Garuda* to the dock, I focused on something that *had* changed. Gistok, now scurrying toward us, was carrying another orangutan. What a sight! Riska and I began asking rapid questions — Who, how, when? — as the two orangutans came tumbling over the dock and scrambled underneath to peer through the planks at us. Twenty long fingers reached out, and instantly enchanted by this impish piece of mischief and his new little mate, I thought, Gistok has been tamed!

This trip is going to be different, thought I.

Only later, when Gistok dragged me bodily down the path, then fell over backward, pulling me down on my face, or when he tucked one long arm around a tree trunk and levered me onto my side, or when he rubbed his lips over my hand so that I could feel his large teeth scrape against my skin, did I admit my mistake. Whatever expectations I'd had about Gistok, and about my reaction to the park that surrounded him, had been premature, at the very least.

Riska and I moved into the cabin like children sent to the backyard with a flashlight, two chairs, and a blanket. We made ourselves a nest. We had enough food to last three weeks. The barrels of river water sat all in a line in the *mandi* although since this is also where Riska left food scraps, visits to the windowless cubicle, which also lacked a drain, became increasingly slippery and pungent. Still, we each had a room, and this was a luxury for Riska that put everything else in perspective. The windows over our beds had one pane of smeary glass each (although they were sealed shut), and there was a small plastic table in each room for a kerosene lantern so that we could read or write our separate ways to sleep. The first night I dreamed that a gang of thugs had broken in and a woman lying next to me on the floor shouted, "Shoot me first!" I awoke at five o'clock hearing the orchestra of birds calling and whistling, an orchestra in which each instrument plays a different tune, and realized that the thugs had been Gistok and his new companion, Boim, thundering over the roof.

In Eden, God set down rules for human beings — we were not to know or understand the mystery of consciousness — but we refused to abstain, and orangutans are refusing, too. Tailless like men, they are stretching the boundaries, trying to stand upright, even starting to think of themselves as individuals who are separate and unique! All day the two ex-captives climbed from one room to another, following us, staring in, so close to the table when we ate that we could see them flattening their comical serious faces hard against the window by our plates. At night they were replaced by mosquitoes and two swallows who flew in and out, resting on the walls precariously as if they were waiting for us to build them a nest as pleasant as the one we'd built for ourselves. In the utter darkness, Riska and I swam through the cabin, a tiny kerosene lantern next to her in the kitchen and a tiny one next to me, so that when I stepped out to talk to her it was into total blindness. I stepped out because I knew how alone she must feel and because I missed my own girls with a terrible ache.

Life and Death at Camp Leakey

Among most Dayak tribes . . . the penalty for making fun, even of fish or frog, is "petrification," turning into eternal stone through hail.

— Tom Harrisson

Because we would have the boat for only three days, we set off the first morning for Camp Leakey, passing a small family kelotok and waving cordially. Then I saw that they were dragging behind them a quarter mile of hand-cut logs. "Will they be rich?" I asked Riska.

"The boss will be rich," she said.

Men spend months in the forests, working alone or in small groups or as part of huge companies, and when the water is high, during the rainy season, they move the timber they've cut inch by painful inch to the river's edge and tie it into a long raft. A good deal of this log traffic is illegal, but there are lucrative timber concessions given out by powerful men to their relatives or cronies or to the generals and politicos who have helped them stay powerful. These concessionees may agree to take the burden of illegal logs off a small-time timber poacher's back, so to speak. Or the poacher may be "caught" and forfeit his bundle of valuable logs to the police.

The fact that the Dayak are cutting down the rainforest that supports them is easier to understand when you understand the Indonesian government's transmigration program to encourage thousands of Javanese to move to Kalimantan and take up farming. When they arrive, the Javanese realize their mistake. Sometimes the trees have already been bulldozed. If not, the Javanese can only keep a piece of land if they clear it. In either case, without its trees, the land is worthless for farming. The soil is of microscopic depth and, without roots to hold it in place, immediately disappears. This is the reason for swidden agriculture, which furnishes a ladang with enough nutrients to grow a single crop of rice. After a crop or two, the ladang is used for rubber or fruit, and eventually something comes back: second-growth forest. But nowadays, the Dayak must cut trees for cash, because they are being crowded out and forced to live in a cash economy. They cut, not only to clear, but also to sell or to earn a salary from a timber company. A trip up any river reveals terrible blank stretches behind the thinnest pandanus screen.

There are no Dayak on this river. The people who live along its banks are Melayu or Javanese or Sulawesi immigrants, like Yadi's family, who are Bugis. Pak Akhyar, the woodsman Biruté had employed for so many years, is Dayak, but from a distant village, and the rest of her staff are from Pasir Panjang, the landlocked town where she lives.

"Pak Akhyar won't be here," Riska said, as we rounded the second bend in the river. "He's been sick. Something wrong with his eye. We haven't seen him since you left."

Indeed, Camp Leakey was mysteriously quiet — there was not a soul around when we arrived, not even Tom, the orangutan who used to hang out on the old ranger tower surveying kelotoks and human visitors. The tower was deserted and the trees were deserted, too, and the long, hot dock was longer and hotter than ever, while the sky was an ominous gray, intensifying the emptiness. On the river, there were no other boats, and there were no voices, no radios playing in the camp. There was a sense, almost, of disaster, except that all we were missing was the bustle of human beings and their unnatural relations with a group of animals who should have been wild.

Maybe everyone had gone back to the forest.

Wearing long pants and long-sleeved shirts, Riska and I set out with cameras and two bottles of water. The temperature was hovering around 44°C — which I translated to 112°F — but somewhere on the path ahead, leaning into the heat and deadly haze, I thought I saw Pak Akhyar standing between two trees. The ghost looked so much like the man who had walked us through the forest six months before that I caught my breath, thinking, If anything happened to Pak Akhyar, he'd haunt this place . . . But Riska saw him, too, and greeted him respectfully, as if he had returned in the flesh.

When he spoke, she made the noise in her throat that is an unconscious sound of assent — "ea ea ea" — and turned to me. "He is saying that six months ago a cat died here after eating a poisonous lizard."

I said, "Yes. We were here then! I remember that."

"And he says he picked up the cat and he wrapped it in an empty rice sack . . ."

"After we left . . . after we showed it to him . . ."

"And then he dug a hole in the ground and finally he buried the cat like that. And the next morning when he opened his eyes, one of them was very painful, as if pins were sticking in it . . ."

Riska paused and turned back to the thin apparition, who was lighting one of the cigarettes we had offered and speaking so softly that his voice was only wind blowing the leaves. "So he waited a few days for a boat ride down to Kumai, and he went to the clinic and the doctor gave him medicine. He says he is not a rich man and this medicine was very expensive and it took a long time for the treatment, but he says the eye was not better. The treatment cost eight hundred dollars in American money." There was another hushed explanation, all in tones of gray. This was almost a year's salary.

"At the end of it, his sight was almost gone and the eye was still inflamed so Pak Akhyar left Kumai and went upriver to his village to see the medicine man. When he arrived there, he found that his wife was very sick. The medicine man said she was sick because she had a very bad magic put on her from someone who was angry, and then the medicine man said he could fight it but he did not want to do that because it would cost a lot of his energy and that the magic would have to run its course. Then he saw that Pak Akhyar was blind in one eye because he had broken a strong taboo. He had buried an animal in the cloth that is meant to cover rice, which is sacred to us. The medicine man treated him with water he had spoken to and held the flowers over Pak Akhyar but time passed and the eye was not much better, so the medicine man said the rice sack was interfering with the cure and told him to come back to this place and dig up the cat and unwrap it and so he has just arrived here."

Pak Akhyar had not yet begun the task of exhuming the cat. He had arrived only an hour ahead of us and he was preparing himself. He knew very well where the magic against his wife came from, he said. A relative was angry because Pak Akhyar found him tapping their rubber trees and made him stop. This relative had put the bad magic on his wife and made her sick. But the eye was something else.

It seemed significant to me that I had found the dead cat on my earlier visit to Camp Leakey and that now Pak Akhyar and I had returned on the same day. I felt very much implicated, very bound up in this story, and I wanted to help. It made me miserable to think of this excellent man of the forest having only one working eye. Even as we stood talking, he held a hand over the left side of his face, as if in pain. I thought it surprising that Biruté, who had employed this man for seventeen years, hadn't sent him to a better clinic. In her own Tanjung Puting brochure she states firmly that serious medical conditions must be treated in Jakarta or Singapore. Surely the cost of a ticket, if not by plane then by ship, would not bankrupt the illustrious OFI, which is supposed to be funding her research, given that Pak Akhyar is possibly the only reliable assistant she has in her employ.

Back on the dock, Riska and I discussed his plight. The beliefs of their ancestors conflict with her own Christianity and with Pak Akhyar's new religion — he converted to Islam after marrying a Muslim woman from Sabah — but in most cases the Dayak are able to manage two systems of thought simultaneously. Some babies, for example, are born with a sign between their eyes that the medicine man can see — a thin bloodline that means the person will be killed by a crocodile. In this case, according to Riska, the only hope is to have the bloodline or vein cut in a ceremony. It can be removed with needles, but if it isn't removed it will grow around one eye and descend and by the time it reaches the mouth, it will be too late to save the child from its fate. Riska's mother told her about a father who refused to acknowledge the seriousness of the mark on his son, although as the line descended he took the boy to the mountains, far away from any river. One day the boy asked his mother to make him a paper crocodile and the mother made him a pretty paper toy and it ate him alive.

"What were the flowers for?" I asked.

Riska said, "To cure his eye. We use seven different flowers depending on the ceremony. Different colors. They're offerings."

I considered my responsibilities to a man I knew only tangentially and with whom I could not speak directly, remembering the dead yellow cat very clearly, as if it were still lying brutally exposed on the sandy path. It might be possible to intervene, but what then? The sky was silvery and the river was heavy and silken. Away in the distance I heard a great call, like the sound of a whale. Riska said it was Kosasih chasing Bagong out of the trees. Two males in contest. Whatever the inspiration, it was the first orangutan "long call" of my life, the call that Biruté had compared, in her Long Beach lecture, to the sound of a drunken elephant. Great Kosasih, adopted child of Pak Akhyar, was calling. At that moment, Riska and I were alone except for Jane Siberry and k.d. lang, whose otherworldly voices mingled in the air with the eerie long call of a male orangutan. Yadi must have been playing a tape on the *Garuda*, but he had disappeared. The boat itself moved restlessly in the water, "Calling All Angels" as it swayed.

In a matter of days, Biruté would be here

again. Plenty of people had told me about her tests of faith, and I was sure I had proved my seriousness. Now we would talk. We would discuss everything from her problems with permits to Pak Akhyar's eye. Riska and I sat on the dock, listening to the blended voices of the river, talking of crocodiles and fate. Later we walked into sheets of rain to see a feeding on the broken footbridge behind the rangers' cabin. I decided to leave my camera behind, which guaranteed the perfect sight of Siswi and Princess and their young ones sitting at equal intervals along the bridge. And after the bananas were distributed, there was Siswi laying her half-blind child on his back on the bridge and lovingly kissing him top to bottom, then taking his tiny penis in her nozzlelike mouth and pulling and sucking on it, while her son lay chewing his bananas in obvious bliss. Siswi was the first orangutan born to an ex-captive at Camp Leakey. Her mother was a great favorite of Biruté's named Siswoyo. Some say her two-and-a-half-year-old baby was blinded in one eye when a ranger threw something at him. Others insist that baby Selamat simply slipped on a dangerous branch.

After the feeding, we followed Princess and Peta and Pangeran (ex-captive offspring are named according to the mother's first initial) up the path, mother and older daughter walking identically, one the longer shadow of the other, while the nine-month-old baby clung to his mother's back. Her reach backward to adjust the baby with a long arm and huge hand was as casual as the reach of a busy mother in the city. Her shamble, since the feet point inward, rocked the baby comfortingly as he held on to her long hair with the grasping instinct of all primate infants.

Biruté decided a long time ago that we were wrong to assume that wild orangutans are entirely arboreal. They come to the ground for some kinds of food — termites, ants, seeds — but they're never graceful down here, while in the trees they move like proverbial gazelles. Nevertheless, Princess and Peta, never wild, chose to walk back to camp, while Siswi and Selamat rode the branches of trees. It was amazing to watch that ride, and that's just what it is. A branch is tested for strength, since an adult orangutan is big and a fall means serious injury or death. Therefore skillful clambering, as it has come to be called, is a necessity — but a necessity that is *learned.*

Learning is cultural if it is passed socially from one generation to the next and is specific to certain groups, not to an entire species. Clambering, like other skills, is partly learned from the mother, who carries her youngster over frightening abysses and thereby teaches him to gauge weight, pliability, and distance. Of course, she also places him in "learning" trees where he can begin to make judgments for himself. Two researchers studying the locomotion of orangutans in Sumatra believe clambering is the reason our common ancestor developed a concept of the self. This would explain the flaw in the socialization theory of intelligence, which has it that we are smart because we evolved in complex social groups and ignores the "solitary" status of orangutans. The researchers, Daniel Povinelli and John Cant, think we learn through failure. The clambering of a large, heavy primate in enormously high trees involves a learned awareness of the self as a causal agent. Over generations, this would encourage the natural selection of a primate with a large neocortex, which would have stood us in good stead when we came out of the trees.

On our way back to the *Garuda*, we found Dr. Muin with a young woman in one of the ubiquitous Indonesian uniforms that looked like every other ubiquitous Indonesian uniform but which was meant to set her apart as a nurse. They had come on an unpleasant errand. On Sunday, Kuspati, the twenty-year-old mother of a nine-month-old baby, had been found dead not far from the dining hall. Dr. Muin had treated her the previous Friday and noticed that her breath was bad. He assumed that she was suffering from a bacterial infection and treated her with antibiotics, giving the baby an injection of vitamins. Earlier the PHPA had run TB tests on seven of the ex-captives at the camp, including Kuspati, but the results weren't available yet. They'd found Kuspati's body on the ground, but the baby had disappeared. "I'm afraid there are so many wild pigs," Dr. Muin said.

He and the woman in uniform had been performing an autopsy on Kuspati, since the concern about tuberculosis was heightened by this unexpected death.

"Would it come from humans?" I asked.

He said, "It's impossible to know. It may even come from the wild population, but it's definitely highly contagious and represents a great threat to the orangutans in the forest."

Riska and I were sitting on the porch of the rangers' cabin watching Princess and Peta and the baby playing together when suddenly Kosasih came rumbling up the path. At that, Princess grabbed the baby and hurriedly clambered up the outside of an old cage where Biruté used to keep ex-captive babies at night. Peta climbed up after her and they scrambled in under the roof and kept out of the huge male ape's sight while he lay down confidently across the path, king of the camp, if not of the forest. Then, as suddenly as he had appeared, he stood up and headed for a tree. Something had disturbed him, something that we couldn't see or hear. Although his usual gait on the ground was lethargic, he proved that he could climb really fast. In moments, he was hand-over-handing and -footing upward, making it to the top of a tall tree just as another rain began. Obviously, he had sensed it coming.

There were the remnants of an old nest in the top of Kosasih's chosen tree, and he began to make a new one just under it, yanking at leafy branches and snapping them off impatiently, padding and padding his nest, feathering it, arranging, snapping, pushing, standing up and leaning out to stamp more and more branches into place. When the destruction of the treetop was almost complete, when finally there was nothing left above him but the umbrella of old nest under which he sat, he arranged himself in the new nest very carefully. The sight of the huge fellow in a small nest at the top of a tall, now nearly branchless, swaying tree was grand and ridiculous, but Kosasih still wasn't satisfied.

The rain had grown heavier and the upper nest simply wasn't providing enough shelter. Now he reached out of the green cone and began snapping off the remaining, lower branches. He was a cone of green leaves with a long brown arm making itself bigger and bigger. It reminded me of my favorite childhood book, *The Bear That Wasn't*, in which the bear wakes up in a pile of autumn leaves and goes into a factory, by mistake, to hibernate. "I'm a bear," he keeps telling the workers, but no one believes him. How can he be a bear when he's working in a factory? "You're just a silly man who thinks he's a bear and wears a fur coat," they say. For Kosasih there was a lot of arranging and adjusting to do before he was convinced that he was an orangutan comfortably concealed at the top of a tree.

Although orangutans begin making nests as early as fourteen months old, they must learn the technique just as they learn clambering. Branches must be properly broken and arranged, for eventually they must support the weight of a full-grown orangutan at the top of the canopy, an engineering feat that requires fairly subtle cognitive processes. There has been some disagreement among the specialists about whether nest building is an example of tool use. For reasons that escape me, most scientists define tools as things that effect change in another object, and though orangutans use branches to scare away people, to fan themselves, to make nests, to provide shelter from rain, and to clean themselves, none of those actions is quite like the use a chimpanzee makes of a branch when she fishes for termites.

When the rain stopped, I looked over at the old orangutan cage, where Princess was hiding. All I could see of her was one red-brown hand, and I swear it was holding a rice sack over the opening to keep the rain off herself and her children — "more nearly a mechanical genius," in R. M. Yerkes's words.

Riska

Narrative language use would seem intimately tied to personal memories.

— Daniel Hart

It was definitely more pleasant to be at Camp Leakey than at our station, where we were Gistok's prisoners. At Camp Leakey it was intensely hot, but there were shaded paths and well-behaved orangutans and water in which it was safe to swim and bathe. At Tanjung Harapan there was only the bolted-shut house and, outside, the hot, unrelenting sun and Gistok's relentless demands.

At Camp Leakey, after our hike, Riska and I went to the rangers' cabin, establishing a pattern for the afternoons to come. We'd sit on the floor or on one of the three broken rattan chairs, drinking a warm Coke from the sales shelf and hearing whatever news the boys had picked up on the CB. Three kittens and a mother cat could be visited in the room where the boys slept, and sometimes, in the doorway, there was another cat — a clouded leopard — but it was a toy, left by a Japanese tourist. The boys used it to scare the orangutans. Although none of the ex-captives, as far as we know, has ever seen one in the flesh, clouded leopards are the natural predators of orangutans, and with the decoy in the doorway, no orangutan would enter. This is true "hard wiring," as opposed to cultural learning.

When we had rested enough to make it back down to the dock, we might find Pak Akhyar and the rangers and boat crews bathing and washing their clothes. Siswi and her one-eyed baby might be sitting on the dock doing much the same thing. Orangutans love soap, and Siswi rubs it up and down on her arms, making suds she can spread around and then lick. First a little soap, then a dribble of water, then some rubbing, then more water, then a taste and some for the baby, who does exactly the same thing. Watching Siswi and Selamat wash is a perfect chance to see a cultural practice in action, this one passed from human to orangutan to orangutan.

Then Riska and I would join the boys in the water, grateful for the relief of its chill, for its lack of mercury (there are no mines on this branch), and its apparent lack of crocodiles. We'd have washed our clothes at lunchtime and hung them on the dock railing or laid them flat on the boat's top deck, and we swam in the clothes on our backs, washing ourselves and our hair. Afterward, we changed into the dry clothes modestly, a little at a time, slipping something off and something on so as not to expose ourselves, and laying the clothes we had swum in out to dry.

By then it would be late afternoon, when the park is officially closed but no one would be around to complain. Siswi would cradle her clean child. The boys would smoke quietly on the dock. Pak Akhyar would finish his bathing and go up to his meal. We would sit on the tip of a long dock dangling our legs in the Sekonyer River and watching the light change. We'd have a cup of tea on the way home as it grew dark. We'd lie on the top deck and look up to see if the rain had passed for good and close our eyes and try to smell the scents that come with the wind. The light from the boat would throw its arc into hidden green places, and the sound of the motor would ricochet off a thousand secrets. We'd tune our senses, we'd listen, we'd project ourselves into the surrounding trees.

One morning, my whole face swollen from sunburn, my nose purple, I remembered the warning on the bottle of antimalaria pills — "causes extreme sun-sensitivity" — and of course I felt stupid. I'd brought a hat that was designed for rain and was too hot to wear. I'd brought a fleece jacket, not knowing that Riska's idea of a chill was anything under 40°C! And I'd forgotten my sunscreen. Along with my swollen face, I was covered with itchy welts again, especially on my legs. Now Riska decided that I was allergic, not to the water, but to the grass!

It was Valentine's Day. I opened the cards I'd brought with me to ward off loneliness and a present Esta had stuck in my bag. The present was a wind-up toy called Bambi that had detachable plastic antlers and a little key. Once wound, Bambi managed to hop around on our linoleum floor, which was better than I could do. I was too miserable to do anything but sit in one of the same plastic chairs I had sat in six months before and glare at Gistok through the window. Riska looked at Bambi hop and then at me and then at it again. Esta had sent her a box of heart-shaped chocolates. These we consumed almost immediately. After a hot breakfast of canned fish and vegetables and rice, we cleared the table under Gistok's curious gaze, his long, hairy fingers wrapped tightly around the ventilation boards above the glass, tiny Boim at his side.

Tanjung Harapan is not designed for comfortable hiking. It's a piece of cleared land between river and forest where once there was a village and where now there is only this collection of cabins and the unvisited graves. The rangers' cabin was inhabited by Alan, who had been here before, and a ranger who was new to me. Another cabin was inhabited by a couple of young British scientists. Another was supposed to be a museum. Another was a wreck. The last one was ours, where we were to live for three weeks. Down a path and well out of sight behind trees was the new PHPA clinic, a spanking-white building, so modern in aspect that Astra, who was still living inside, could only be seen behind the bars of a high window.

We'd purchased two tapes in Pangkalan Bun at Riska's request: an Indonesian rendition of country music and a pirated Elvis. Unless we started talking — fast — it was going to be hours every day of imitation John Denver on Riska's Walkman, "Blue Suede Shoes," or Gistok.

I had finished *Uncle Tom's Cabin* and brought it out of my room. What would my guide make of the notion of slavery? ("Why are they so popular?" she had asked once, when we'd seen a black entertainer on TV, and she was enormously conscious of the shadings among her own people.) "I've been reading about a time when things were terrible," I told her. "There was a civil war."

"I'd like to read it."

"Sure. Okay."

The trouble was that Riska saw me as a client or "guest," as she prefers to say, whereas I wanted more and more to be her friend.

I am her mother's age, so there was that difference to deal with, along with culture and background and all the rest. She lives with a foot planted in two cultures — traditional Dayak and modern Indonesian. As a Christian and a Dayak, she's an outsider among Muslims, and as an urban woman she can't really go back to the forest. Riska is single, independent, and the only Dayak woman in Kalimantan who is an official guide. "Have you been writing?" I asked her, from my red plastic chair.

She showed me the thin, lined notebook.

"You're still writing in English? Why're you doing that?"

"So you can read it."

"But . . ."

"It doesn't matter. Nobody else would be interested anyway."

"I wouldn't be so sure. You've had an amazing life."

"You don't know the real me," she told me then, on the morning of the sunburn. "You might not like me if you did."

"I'd like you better," I said, but I didn't offer her any proof. We were two women sharing a tiny house in the forest. Or jungle, as she would say. I depended on her temporarily, but the chasm that separated us in terms of experience and expectations was so wide that I couldn't decide how to bridge it. I looked at her notes. She had written in great detail about the *tiwah,* or funeral practices, of the Tumon Dayak. An anthropologist would have been overjoyed — there were technical terms, recipes, tribal practices. The hot morning poured on. Gistok and Boim at the glass.

Amazed by Riska's meticulous descriptions, and by the care she had taken to get things right, I read every word. The creator of Riska's people has two aspects: hornbill and crocodile, male and female, of sky and of water. This dual deity gave the Dayak (of which the Tumon are a tiny part) their laws and customs, including headhunting and rice growing, which are associated, respectively, with males and females. There are ancestor spirits, once mortal, who now occupy the landscape. They're contacted through dreams and messengers, usually birds. Riska is Christian, but she is conditioned by her culture to feel the sentience of rice as well as the very ground in which it grows. I put the notebook down. "What was it like being in the village?" I asked.

"My village?"

I said, "You want to tell the story of the Dayak, but you're part of it, aren't you?" I'd learned about the carving of the coffin (hornbills and crocodiles), the draining of the corpse, the carving of posts. "Tell me about a funeral you remember," I suggested. "Tell it through your own eyes, the way it felt. If you're going to write, you need a point of view. You need to expose yourself."

By mid-afternoon, Riska had been pulled back to her childhood enough that it was clear she had paid a price for the privilege of living in two worlds. "There are things in my life I can't say." There were caverns, places too dark or too private to go. But I'd brought some blank tapes and we tried using the tape recorder for a kind of interview, even though she was frustrated by my efforts to dig and at one point was close to tears. What was I getting at? Why did I care?

"I'm going for a sleep," she said finally, getting up, pushing her chair back, although the idea of going to the bed afloat in my own unventilated room was unthinkable. I sat looking at the tape recorder and thought of the long days ahead of us without even the *Garuda* to carry us upstream. Yadi had gone back to Kumai. He was waiting for Biruté and her movie star. What was I doing in Borneo, where it was too hot to think, too hot to move? Where my face hurt from the inside out. Where an open door and a little breeze would have been the most welcome thing in the world, except that it would expose me to ex-captive, marauding orangutans?

At four o'clock, it begins to be cool enough that movement is just bearable again. Between five and seven, the light gets longer and softer and there are shadows on surfaces. Greens deepen. The sky becomes opaque. That afternoon, as the hard heat began to lift, I stood, wet with sweat, on the porch and tried to throw off the oppression of the closed and silent rooms. As if the effort of bringing back her childhood had been physically exhausting, Riska was sound asleep. I took a deep breath. Yes, it was ever so slightly cooler and the sunlight didn't hurt my skin. A pair of kingfishers flew past! The smell of a distant cooking fire rose up from across the river and mixed with a nearer, marshier scent. My forest ancestors would have recognized every detail of those smells, would have absorbed them the way I absorb the particular flavors of a soup. This time of day would have furnished them special delights, which is why I had come, after all — to waken those senses and find my most elemental self. Locking the door against

Gistok, I threw the key in over the transom, so that Riska would not be trapped inside.

Locked out, locked away from everything even faintly familiar — my books, clothes, comforts, my language, even my guide — I was staring at a river full of crocodiles. I was allergic to the grass. I had no boat. No phone. No mailbox. No one could contact me even if they tried. In a moment I would walk into the forest and if I didn't come back, no one would know where to look for my remains. I'd kicked off the traces; that's how I felt. For almost thirty years I'd been responsible for children. And to them. In fact, I'd been answerable to someone all my life. To be sensible, to be respectable, to stay within the law, within the bounds of etiquette, inside well-marked cultural lines. Like the orangutans, I was a creature of culture. And like Riska, only she'd gone a lot further from her origins than I. From the start, I'd been so closely monitored that when I jumped, I never knew whether it was a reaction to the lines I was crossing or a response to some part of my true self. I couldn't think of a single minute of my life that hadn't belonged to someone else. Even now, on the other side of the locked door, there was Riska. Maybe I should stay unattached, forget any personal feelings for her. I'd come from another world and I had no right, or duty even, to make her part of mine. Writing the truth, about herself or anything else, would permanently change her life. I had locked the door. Fine. And when I looked around for

Gistok, he was out of sight.

I thought, If fear is connected to other people's claims, then being invisible I should be fearless. And being fearless, I'm free. "Out of here," as my kids would say. Four hours' walk away there was a piece of land held by Birutė's husband, Pak Bohap. Anyway, it was ostensibly in his name. I'd heard about it from the rangers, from other visitors, and from people in Pangkalan Bun and Kumai. Birutė's secret forest. I'd heard that Birutė kept a group of ex-captive orangutans on this land under the care of some very isolated workers who were extremely unfriendly. Even hostile. I'd been told that, denied access to Camp Leakey, she protected this secret place as vigilantly as we protect our secret selves. There were guards. It was "private property." Very dangerous to trespass.

There were stories from former volunteers and associates, and also locals. They were never denied.

I kept the river on my right and the trees on my left. The path was sandy and I searched it for fire ants. Nothing. Nobody. That was what I felt. Around me, beneath and above, nothingness. Even meeting up with a wild pig or a poisonous snake would have been less terrible than . . . And what if a snake did come slithering out of the grass? I was only wearing my Tevas and I hadn't told Alan I was going out of camp. I'd broken a cardinal rule: *Never go out alone and always notify* . . . If something happened, I would cause trouble for him. It was darkening, and that happens fast. If I went any farther, it would be too dark to find my way back. Trees on my left, gathering shadows like birds, river on my right, lungs out of breath, feet pounding the earth, I ran toward that hidden place in the woods. Passed the graveyard. Got to the old, wobbly tower. I was picking up speed, watching my feet grip my sandals, watching the ground move under me.

In front of me, Birutė's secret place. Behind me, Riska's hidden life.

Birutė says we are social animals. She says we're social because of our greed. But maybe it's something else. If clambering taught us to think of ourselves as separate and unique, as causal agents, it made us aware of ourselves and our movements so that we could paint on the walls of caves when we came down to the ground. What we painted was stories and images made for another set of eyes, a spirit or god or fellow being. It's the awareness of ourselves amongst others that causes us to create.

I decided to ignore the secret forest and turn around. I wanted my lover, my children, the friends I've adopted for keeps. And my work. Words. Riska. The door to the cabin was still locked, so I turned toward the river, where I could sit on the dock and cool my feet. It was dusk. Riska would soon wake up and light our lamp. I neared the dock and saw something strange. Gistok was sitting in the river, quietly splashing himself with a meditative flick of his wrist. Slowly, as if absorbed in the deepest thought, he examined his fingers, trailing them through the current so he could study the drops that fell when he lifted them. Delighted by the sight of his toes underwater, he grinned at them and then looked up at me and grinned again.

Orangutans like sitting on docks and playing with soap, but they never swim. Still, Gistok sat in the river. He was alone. I got in with him. Companionably, we watched our hands waver under the surface, both of us wordless, made of the same sensibilities. Each of us with a brain that creates images and the senses that feed them joined in the realm of gestures, expressions and empathy, sweetly communicating. No tugging this time. No overpowering strength. The sky reddened briefly, and Gistok leaned over to peer at some passing fish. If I'd had a pair of goggles, I'd have given them to him.

That night Riska and I ate our meal and prepared ourselves for sleep. Mosquito coils, lanterns turned down, teeth brushed, dishes scraped.

"There's something I think I should tell you," she said in the wavering light. "Did you know I have a little daughter?"

I shook my head.

"I didn't think so. It was bothering me."

I do like you better, I wanted to say, but I couldn't find the words to explain. What draws us close is not just sharing the everyday things. There are also frightening and terrible truths.

How much was I giving of myself? I couldn't even tell her how I felt.

I went to sleep and dreamed of my mother visiting me in a house that was tall and white. We walked around on its green lawn, and she told me the name of her favorite rose and gave me instructions about planting it. I tried to explain that I would never live long enough in that house to see a rose come to bloom, but suddenly she fell on the grass and I picked her up and carried her on around the white house, feeling such tenderness for her that all the next day, when my senses were again awake, the feeling persisted, the weight of my mother in my arms.

Pak Herry

Late one night I went down to the dock and sat with Alan, the ranger, and his brother Supri, who used to work for the PHPA at Camp Leakey and now worked at the second station. He had come down by sampan to visit, and since neither of the boys spoke English, we sat in silence on the wooden dock, swatting mosquitoes and looking around at the bowl of sky over us that came down on all sides. The boys smoked peacefully in their sarongs while the astonishing stars came out and I wished I could ask them some questions. I knew Supri had been at Camp Leakey when Augustin and Purwasih arrived, the first two of the Twisted Sisters. Biruté had a Canadian assistant named Michele, and she had wanted to give one of the new orangutans that name but the PHPA had something else in mind. They put Supri in charge of the new ex-captives instead of handing them over to Biruté and named them Augustin and Purwasih. Until then, Biruté and the OFI had always named the ex-captives, and even now they refuse to acknowledge any PHPA names for the orangutans. Nevertheless, slowly but surely, control was being wrested from Biruté's hands.

I also knew that Supri had been working at the

second station when a young orangutan named Somalia had been taken out of his arms by Biruté and carried downriver to her house in Pasir Panjang. I'd read about Somalia on an Internet site posted by the OFI and in the OFI newsletter.

"Rescued by Dr. Galdikas only a few days from death, Somalia was, literally, skin and bones when I first held him," wrote a volunteer. "Even after a month of loving (and constant) care, he weighed less than ten pounds, and clung ferociously to his foster moms."

I was longing to ask Supri about all this, but because of our lack of shared language, my questions would have to wait.

All day we had expected Dr. Muin to arrive with the four new orangutans, but he hadn't come. When a boat showed up carrying tourists, I felt so reclusive that I stayed inside while Riska went out to talk. Actually, it was more than reclusive — this thing I felt. I'd become territorial, even defensive, avoiding contact with the British scientists in the next cabin, who were working on a project that had something to do with tropical plants. I disliked it when any boats turned up. Small talk made me squirm because I was deep in my own thoughts. The tourists turned out to be German, and one of them was badly injured when Gistok pulled him down hard on the dock, but I didn't have much sympathy for the German. Maybe Gistok was territorial, too.

Twice while Riska and I were out in a sampan,

we had come back to find that he had broken into our cabin by smashing windows and had run off with food and our personal belongings. He'd taken Riska's lipstick up a tree. He'd taken our precious watermelon. He'd eaten all of our fruit and most of our rice before he left, leaving bits of glass behind. He'd broken three plates and a bottle of lemon drink. These were losses of the most intense magnitude. A bottle of antibiotics was gone, though there were a few empty capsules stuck together in the grass. After his sins, Gistok had applied the lipstick to his mouth and hung from a tree by the porch in order that we could admire his new countenance. Instead, we had locked the door, knowing it might not do any good, and counted our change.

Eventually the new orangutans arrived. They came up the river in a speedboat with Dr. Muin and Pak Herry, a man I had still not yet met. I took a good look at him while the wooden crate was being opened with hammers and crowbars, and decided I was going to like him. His is an open, entirely friendly face, and he's plump in a nice middle-aged way. Also, he's intelligent and funny; I could tell that straight away. He was taking pictures and cracking jokes and at the same time managing to lend a hand with opening the crate and carefully bringing out the frightened orangutans. He was treating the young men under his supervision and the young orangutans under theirs with respect. Most surprising, he seemed neither alarmed nor impressed to find a North American writer at his research station. Pak Herry, otherwise known as Dr. Herry Susilo, had studied at Michigan State University ("I am a Spartan!"). That meant open debate had to be at least as familiar to him as snow and ice.

Dr. Muin was all smiles, too, and obviously proud of his new patients, who seemed puny and terrified. One, though, had hair so long and silky that it was hard to keep my hands off him. What a beauty! We oohed and aahed as he was carried into the clinic for a Dettol bath. Among the four, there was only one female, and she cried and squeaked piteously, as if her small heart was about to break. Those young orangutans had been pulled away from their dead mothers first and then pulled away from the humans they'd been depending on. They'd had two days of traveling in small containers, in addition to vaccinations and other unpleasantness, and they were ready for a rest and some serious comforting. Or so I would have thought.

Pak Herry told me about his trip to Taiwan with Willie Smits and the head of the PHPA. "There are two hundred orangutans there," he said, looking troubled by this ominous waiting list, "and where in the world could we put them even if we brought them in? There just isn't the space. We have three missions here. First to save, then to study, then to make use of the park. But we are always stuck on 'save.' Do you know, last December I stopped a hundred and thirty-seven men from cutting lumber in the park. And that was in just one week!"

"What did you do with them?"

"Gave them a lecture and confiscated their saws."

"I hear there may be TB at Camp Leakey."

"Well, we ran some tests."

"Is that why Kuspati died?"

Pak Herry looked at Dr. Muin, who nodded. "But what do we do with them?" Again, he threw up his hands. "Put them to sleep? There would be a terrible fuss about that! Can you imagine? We are concerned about preserving the species, but some people are more concerned about animal rights. I mean, rights for the individual."

"Maybe you should open a sanatorium," I said, half in jest.

"At Smits's place," he told me, "they have all the TB patients in a separate cage."

"And their care is very expensive," Dr. Muin put in. "The new law that no ex-captives can go into an area with wild orangutans is very good, but it means we must find new places to release them, and there are not many places left."

Pak Herry said, "We told Taiwan they must check for TB before we accept any of their orangutans. And we will take none over seven or eight years old. By then it's too late to go back to the wild. Like Gistok. I tried to find a place for him. The Safari Park or the zoo in Jakarta. But they have too many orangutans. They don't want him. Poor Gistok. I would like to spend my time making a good information center and many things, but I am too busy trying to save."

I remembered Carey and Trevor's enthusiastic praise of Pak Herry, the tales they had told of his efforts to rid the park of poachers, often taking off by himself in a speedboat to track them down. Then I remembered the illegal orangutans at Biruté's house. If poachers were no match for him, apparently Biruté was. We were sitting in the white-washed clinic, which smelled strongly of the Dettol baths that were being administered to the new ex-captives, and Pak Herry began to describe in detail the center he imagined here at Tanjung Harapan. It would have scientific exhibits so visitors would learn about orangutans before they came to the park. The museum would officially open.

"And we now have a skeleton!" said the doctor.

"So we would like to — how do you say — reconstitute her."

"Who? Kuspati?"

"Yes."

"But you buried her."

Dr. Muin's eyes twinkled.

"A digging-up ceremony? Like the Dayak have?"

Pak Herry said, "No ceremony! Although we must wait two months for the flesh to fall away, so it is very much the same. Once before we thought we had a skeleton. Biruté's workers said they'd found a dead wild orangutan at Camp

Leakey. So I asked them to show me where it was buried. They took me up there and showed me a grave, and when we dug it up it was a wild pig! Later, I went back. They showed me another grave. Of Siswoyo. So I took her out. She was in good condition because Biruté had wrapped her carefully. But that night she came to my house and she was crying. She begged me to put Siswoyo's body back in her grave and told me Siswoyo was very special to her. They had a very special relationship. Siswoyo was the first orangutan she raised. She told me the story. So I took Siswoyo back, and later I learned that she has the bones of the wild orangutan, herself. Biruté. She told me she would let me have them, but she never has and I won't ask her again. I don't feel comfortable at Camp Leakey. Even though I must go there; it is my responsibility. And even though she is not there."

He shook his head. "Putting together a skeleton will be very, very difficult. Especially the hand and wrist. So many bones there. But I have found a good glue for this purpose and if we could find an artificial human skeleton . . . Isn't it so, Muin? It's only to show the similarities."

I remembered the story Biruté had written about Siswoyo, who died during childbirth complications just after Biruté herself had suffered a miscarriage. It seemed, what with one thing and another, with Biruté's sentiments and Pak Akhyar's taboos, with human deviousness and scientific inquisitiveness, I was in the middle of a jungle of expressions and gestures that were no longer being mutually interpreted.

For another two nights, I dreamed of houses. They were wildly decorated. They had rooms with glass walls. There were upstairs floors of ice and downstairs floors of grass. I was dazzled by my unconscious inventiveness, but the source was obvious. We had left four shivering orangutans locked in a tropical clinic with poor little Astra and a pile of small bananas. All over Indonesia, owning an orangutan is illegal, but the habit persists. Having an orangutan in your house is prestigious; it's something the colonials made popular. Baby orangutans are often taken from the arms of wealthy businessmen and army generals. They're fun, and then they are no longer fun. And then they are dangerous.

The second night I woke to find that my lantern had gone out — not a problem, except it was impossible to find my way to the toilet and then find the right place to pee. The dark at the edge of the trees is solid, but under the stars it liquefies. I decided to unlock the door and step out. Gistok would be asleep. That much was fine. But the air was swimming with mosquitoes, and even so, I crouched. Swatting at the air, I listened for the faint sound of orangutans weeping from the clinic a few yards away and thought, Surely they shouldn't be in there without a warm body to cling to. If I'd had a key, I'd have gone on through the darkness to check on their sleep. Instead, I exposed my bare backside to the knee-high grass. Back in the cabin, covered with bites and already experiencing an allergic reaction, I felt around in the dark for the first-aid cream, grabbed the wrong tube and rubbed toothpaste all over my bottom. There should be a parable somewhere in this. But it eludes me.

Mr. Ralph

Most of the naive orangutans soon adopted the strategy of following someone else through the forest.

— Anne Russon

For a year, I'd been hearing stories about a man in his seventies who has followed Biruté around for ten or more years. "You must call him," people had told me. "He knows everything." Mr. Ralph, as he is affectionately known, lives in New Jersey, and he sounds as if he lives in New Jersey, but that part of his life doesn't count for much with him. He made his first trip to Indonesia when he was fifty-two, and he's been going back once, twice, three times a year ever since. When someone mentioned that he was staying across the river at the Rimba Lodge, I decided to go over. We'd been stuck in our rut at Tanjung Harapan, but Riska had unexpectedly managed to hire the *Garuda III* for two days, driven by Yadi's brother, Yatno, and I thought Mr. Ralph, who had no transport of his own, might like to ride up to Camp Leakey with us, especially since Biruté and her movie star were now overdue.

The big kelotok wouldn't arrive until the next

morning, so we crossed the river in the wooden sampan Tanjung Harapan keeps sunk by its dock. Before we paddled, we had to lift the boat gradually out of the water, me grabbing one end, and Riska the other. Pulling a boat full of water out of a river is a tricky job, but as usual, Riska had a method — pulling, tilting, and shaking, while at the same time rocking the boat from one side to the other. Then we bailed.

All this time Mr. Ralph sat alone at a table in the Rimba dining room, unaware of our efforts. He is a dapper, trimly bearded man with a slight hunch, no doubt because he lugs a huge bag of camera supplies everywhere he goes. "Are you the famous Mr. Ralph?" I asked, when I threw back the doors of the otherwise empty room.

He lifted his head, then lowered it again. "Why do you ask?"

"We have mutual friends. In Toronto. I've been instructed to look out for you."

"In that case, sit down."

He was eating a plate of *nasi goreng,* and something about the way he was staring at his plate and at the checkered tablecloth and the glass of beer told me Mr. Ralph was a loner, but also lonely; all the more reason to announce to this stranger that I had secured a boat for the next day and wondered if he'd like to ride up to Camp Leakey with us.

He looked up. The glance got as far as my chin. "Sure. What time? There are two young American ladies playing Ping-Pong in there" —

he gestured at the room behind him — "who are also looking for a lift."

The two young ladies were Kate and Amy. They had jobs in Jakarta, and they'd come to Borneo to see orangutans. They'd got as far as the Rimba Lodge and now there weren't any boats to take them farther. At this rate they'd go back to Jakarta having seen only Gistok, but they were making the best of it by playing Ping-Pong with a couple of young men on a chartered tour. Borneo has this in common with the nineteenth century: fellow travelers provide whatever kind of fun is to be had. You're stuck at the lodge with nothing to do and no place to go. It's raining. Then someone approaches with an offer of Ping-Pong or an invitation to ride on a boat, and you're suddenly having an adventure and making a friend for life.

I didn't talk to Kate and Amy a whole lot the next day because I was busy adjusting to the bigger, double-motored *Garuda III* and busy talking to Mr. Ralph, who is active with the Los Angeles OFI and refers to Biruté as "the Professor." He told me the OFI is raising money to save the habitat first and, second, to carry on the Professor's research. "We hope to open a clinic," he said.

I said, "But there's a nice new one just over there," and pointed across the river.

He said, "The PHPA can't take care of sick orangutans. They would need a vet."

"They have a vet," I said.

He said, "No, I don't think so."

"So you think the orangutans at Biruté's house are getting better care than the ones in the park?"

"Oh yes, because they have one-on-one in Pasir Panjang. The rangers up here don't know how to take care of babies. We were up here at the second station a couple of years ago and we saw little Somalia up a tree, so weak, so undernourished that the Professor said he would be dead in three days. So we insisted they give him to us and we took him back to the Professor's house."

"And he got better?"

"When he does, we'll put him back in the forest."

"You mean he's not better yet? And there's a vet at her place?"

"No, but there's a swamp, you see, and the orangutans are out there every day. Only at night do they go back to the cage."

I asked Mr. Ralph where Biruté's orangutans come from, and he said, "Well, this is very hush-hush, but the PHPA confiscates a lot of them and sends them to the Professor's house."

"They must have a strange relationship," I said, and Mr. Ralph agreed.

"The original understanding was that she'd take care of the ex-captives. That was stipulated in her permit to study. But all the regulations changed in ninety-two. Before that, anybody could stay at Camp Leakey with her permission.

She built all the buildings there, except for the PHPA station, with her own lumber. But every so often they change the minister of forestry. He appoints the director general of the PHPA, who chooses the head of the park. At one time the head of the park was at odds with the Professor, but he's out now, although the regulations are still in effect. So the Professor is trying to establish a clinic in Pangkalan Bun. We used to have a clinic there, but the PHPA came and took all the orangutans away."

"Maybe because of the Bangkok Six," I suggested.

But according to Mr. Ralph it wasn't the Bangkok Six that changed everything. Like Anne Russon back in Toronto, he believes that Birutė's reputation with the PHPA was damaged by that less-famous group of orangutans — the ten who were sent back to Indonesia from Taiwan.

"The Taiwan OFI sent ten orangutans to the Professor," he told me. "And she sent them to a building outside the zoo in Jakarta, which was rented to the OFI. The orangutans were taken care of by her students. I was with those students. I used to buy the beer and we'd sit around together. Sure, they kidnapped the orangutans to keep the PHPA from taking them, then the police moved in and some of those students ended up in jail. They were even beaten until they gave the orangutans up, which of course they did. And Smits got them. Three were sick. They were sent to Bogor. But seven were sent to

Willie Smits. One died. The OFI didn't even get paid for their care. But Smits was well set up. The Dutch government gave him six million dollars to start his place."

"The Dutch government?"

"Well, the president there, or Prince Bernhard — he's quite a conservationist."

I said I had heard that there are still two hundred orangutans in Taiwan and Mr. Ralph snorted. "Seven hundred, more like it!"

"What's to become of them?"

"They'll have to stay there. They'll grow old and eventually die. Taiwan has stopped the illegal pet trade, but there are plenty of orangutans registered there. And there are plenty more in other places. You can legally keep one as a pet in the United States!"

While the noisy *Garuda III* hurried us upriver, Mr. Ralph went on to tell me how he came to Indonesia in 1971. "We didn't see many animals on that trip," he remembered. "You know, most of them were in cages. But I made a side trip to a Dayak village, and then I spent three years trying to get back. In those days nobody was going to Borneo, but I kept writing to a travel agent here and finally she told me about a group coming from the Netherlands — all kinds of scientists — and eventually I met them in Amsterdam, although later I broke away and got my own guide. Oh, we stopped many places, but finally we found the Dayak and I've been coming back ever since."

The big kelotok's two engines made conversation barely possible. Even so, while Mr. Ralph peered into the passing trees for signs of proboscis and into the sky overhead for kingfishers, I persisted, at times almost screaming, for it seemed to me that if the Movie Star didn't come soon, Mr. Ralph might be as close to the Professor as I was going to get. "If saving the habitat is the main thing," I shouted, "isn't it important to work with the Indonesian government? They're the ones who sell the lumber concessions and control the land. Isn't the constant bickering with the PHPA counterproductive to the interests of the orangutans?"

He blinked. "The government? This president, you know, his brother and his wife are both in the lumber business. She's got part of the plywood mill in Pangkalan Bun! He's honest, maybe, himself, but he looks after the family. All of them. And who knows how long he'll be in power? He's seventy-four! (But then I'm seventy-five!) He could go on a long time. No, we have to save these trees here in the forest where the orangutans live."

"Yes, but these trees are already saved. It's a national park. It's about as safe as it can get."

Another blink. "Well, we're not rich, like so many foundations. We don't have a big benefactor. So it takes a lot of time. Fund-raising, that's the main job. Things were pretty bad here for a while," Mr. Ralph said. "But there's hope. The head of the park and the head of the PHPA and the Minister of Forests — they're all new. So it may get better. Earthwatch may even come back."

"Why did they leave in the first place?"

"Because they couldn't stay up at the camp anymore, and they didn't like that. And they don't like controversy, and there was plenty of that! But the new ones are English. They're not connected to Earthwatch USA."

"You have all these OFI tours doing the research now. Haven't they made up for the loss of Earthwatch?"

"Well, it's not as many as it was, I don't think."

"What does a volunteer do?"

"First he goes out with two assistants and stays to the trails. But once an orangutan is spotted, of course there's no sticking to the trails. It's into the bush. Usually swamp. It's very rough and you can't stop for the night until the orangutan does. Sometimes that means fourteen hours. It's not for an ordinary tourist! It's work. That's why the trips are tax deductible."

"All those hours . . . what are they doing? What kind of research?"

"With the males, we note where they are and what they're doing. But with females, we keep track of all the reactions between mother and child. Or second offspring. We have a regular form to fill out with each and every minute accounted for. Over a fourteen-hour period, that's a lot of notes. Then at the end of the stay the volunteer keeps those notes but makes a final report for the Professor."

"What if the volunteer doesn't know which orangutan she's observing?"

"Then usually the assistant knows, or if not, he runs back to get the Professor and she comes out to identify. These follows start at four A.M. So now the volunteers have to sleep on a boat or get up two hours earlier at the Rimba. It was easier when they stayed at the camp in the team house."

What struck me most about Mr. Ralph was his devotion, as if his pleasure in being on the river had only to do with service, as if the woman was even greater than her cause. Off and on during the day, around Camp Leakey, in the camp, at the feeding, in the dining hall when he was changing film, I kept firing questions at him. He carried a huge lens and a very professional camera as well as a fully equipped, heavy bag. Also a tripod. Flash attachments. His watch registered everything from barometric pressure to mood swings and he reminded me constantly that it was 42 or 43 or 44 degrees. "Barometer rising," he'd say from under his baseball cap or bandana. What he described himself as doing was "waiting for the Professor," who would soon arrive with the Movie Star. "She's going to need me," he'd say.

On the way back to the Rimba he showed me a picture of Kuspati, the orangutan who had died and been given an autopsy. "Taken a week ago," he said. "I guess I got the last picture of her." He keeps a small notebook in one of his oversize pants' pockets and makes a note after taking each photograph. Some of his pictures appear in OFI brochures, but according to Mr. Ralph, he never gets paid for them. "Oh, that would be wrong, since I'm here as a tourist, not on a work permit."

The next morning, Riska and I stopped at the Rimba again on our way upriver to Camp Leakey. "No sign of Birutė?" I asked Mr. Ralph.

"Not yet."

I was waiting, too, sure that once Birutė saw that I had come all the way to Kalimantan, not once but twice, she would relent. At the very least she would know I was serious and speak to me. Everyone else on the river was waiting for the Movie Star. Even among the park rangers there was audible excitement. "They're going to make a movie." "Mrs. Birutė is coming here with a famous movie star." "Julia Roberts." "A movie." Which would prove, once and for all, that Tanjung Puting had a place on the world map. That it was important. That it was glamorous. "I think it's Julia Roberts." "From Hollywood."

It's a big river, but a small society, and foreign visitors are carefully noted. We were also waiting for the arrival of ten Korean businessmen and their hired female companions. Yadi and Yatno and their father had refused to carry this crowd on their boats, not because the men represented

Korindo, the big lumber and plywood company that's clear-cutting Kalimantan's forests, but because the women, being Indonesians, were a disgrace. Still, where there is money, often a way can be found. Someone in Kumai would provide a kelotok. It was all very exciting.

After Mr. Ralph stowed his equipment on board the second morning, and Amy and Kate had taken their favorite positions at the prow, the two-and-a-half-hour ride upriver began in a mood of anticipation. Surely this would be the day. It was a beautiful morning, and I had the unscientific idea that an early start would allow us to stay ahead of the heat. "It doesn't bother you?" I shouted at Mr. Ralph.

"Well, if it does, there's nothing I can do about it."

He wore a baseball cap turned around while he sat on the upper deck explaining, in a very New Jersey voice, why he gives his money directly to Biruté instead of going through the OFI. "That way she doesn't have to pay the taxes," he said, over the roar of the motor. "If you stay at her house, for example, as a volunteer — people are always doing that — if they stay with her, well, she more or less charges for it. She married into a Dayak family and that's expensive. They think all Westerners are rich, so you have to expect . . . She's always hard up for money."

In spite of the massive amounts of camera equipment he carried, I had seen him take no more than two or three pictures. "I'm not interested," he had said on a path at Camp Leakey the day before, seeing gorgeous Yayat lying on his back. "It would be different if he were in a tree." And at the sight of another orangutan and her son and baby silhouetted against the sky, he said, "The sun's not on her face." And about the sun itself, he said, "It's never there when you want it," and about the lack of sun, "I don't like to use flash."

Maybe life is about only two things, safety and excitement, I thought. Traveling, eating, going to work, making love, waiting to take a picture of Biruté. In the waiting itself there was excitement for Mr. Ralph and maybe for all of us. We were past thinking about safety.

I'd decided to take my notebook along with me to the forest, the way Mr. Ralph did, except that mine was large and spiral-bound and inconvenient, impossible to shove into a pocket and too hard to reach tucked away in a pack. Right away, as we neared the camp, I saw a group of orangutans in a tree. The Twisted Sisters were there, a young male named Muchtar, and Siswi, with her half-blind child. All of them were together, more or less clustered in one place, so I set the notebook down and focused my camera at them. In a second what I saw through the lens was my notebook dangling from Siswi's hand high above me. This was the book I'd kept all my notes in. "Siswi, give it back!" I screamed idiotically, since she was making for the highest branches of the tree with baby

Selamat clinging to her side.

Clipped to the notebook was my ballpoint pen, which she uncapped. Next she opened the notebook and turned back a page. I had horrible visions of her dropping my sentences to the bystanders who were already gathering, a cluster of three or four people from another boat, who were now standing beside me, looking up and laughing. I shouted at Riska to run and get candy ("Anything! Anything!") from the rangers' cabin, never mind the rules. "Just get her down!" In fact, I was jumping around at the foot of Siswi's tree, while overhead she made a few tentative marks in the notebook, looked down, and grinned. *Homo sapiens raises arms helplessly.*

Riska came running back with several bananas, and Siswi came down and got close enough to grab one and run back up the tree, holding the notebook out of reach and managing to peel the banana and eat it before she snapped the pen in two and rubbed ink around on the page.

Another grin. She had created a picture.

A small crowd collected on the path, enchanted by her performance. I remember a young couple in shorts who were snickering, but after all I was making a fool of myself while Siswi turned the pages of my spiral-bound book, pursed her lips, and added a few inimitable editorial comments. Then she moved down the trunk a bit so that we could appreciate this, but she stayed just beyond our reach.

What was going to happen next was that Siswi was at last going to go back to the forest. With my notes in hand, she'd have all the information she would ever need in order to be a wild orangutan. She could raise Selamat as he was meant to be raised, for there were plenty of specifics about fruit and nuts and leaves. It was easy enough to imagine the notes being passed around. Tree to tree. Friend to friend. Mom to mom. With Siswi gone, Purwasih and her Twisted Sisters might be next, leaving Camp Leakey like a group of refugees to study Orangutan Ways in the forest.

But I would have nothing to take home with me.

So I made a desperate grab at Selamat's foot, thinking that if I got hold of him, Siswi would have to come down or let go of him.

"Don't touch her baby!" Riska screamed. But it was too late. In a rage of protectiveness, Siswi yanked Selamat and my notebook back up the tree, where she sat glowering from its highest branches.

It would be possible to write the story of the Camp Leakey orangutans as a multigenerational saga. One would have to start with the assumption that the characters were all disturbed, that their behavior would never correspond to the behavior of orangutans raised in the wild. One would admit that the individuals are as biased, cunning, and self-reliant as the usual set of characters in a multigenerational saga. Some are motivated and some are not. Some are complainers and some are not. If Dr. Muin's and Pak Herry's reports are right, there are sixty-one orangutan ex-captives in the national park. Perhaps thirty or so have gone back to the forest, more or less. But forty or so are *adi di sekitas pos,* "hanging around," and that morning Siswi was the star of their show with a brand-new notebook, although she finally came close enough that I could grab it and pull, losing the cover and gaining a piece of orangutan art.

But that wasn't the last drama of the day. In the afternoon, while I was in the dining hall talking to Mr. Ralph and watching him clean a lens, Augustin, with Purwasih on her back and Tata, the third sister, clinging to them, climbed up the window screen. Nothing unusual about this, but Siswi was under the house and she grabbed Augustin's leg and pulled, just as I had grabbed and pulled Selamat's. Her tug caused Augustin to tumble to the ground with Purwasih, and there was a great deal of screaming. Old Tom and young Remaja, hanging around in some nearby trees, were witnesses to the whole thing and expressed themselves angrily with kiss squeaks and mouth-fart noises. What a ruckus! Remaja is a wild orangutan who has decided that life at Camp Leakey can be more fun than life in the forest. She makes me think of Victor, the "wild child" who was raised by Professor Itard in nineteenth-century France. Itard's first aim was to show Victor that life among people was more pleasant than life lived among the "rigours and loneliness" of the forest. But what did Victor respond to? A sudden snowfall brought his most intense expression of joy. He spent many hours staring up at the moon. Not so Remaja. The moon is less interesting than life at Camp Leakey. She would make Itard proud; she goes back and forth between worlds but stares at us and the other ex-captives enviously. She's becoming bicultural — an orangutan who can live in both worlds — but she's doing it in reverse.

The hottest part of that day was spent in the water, and that was the hour the Korean businessmen arrived. The sight of their wretched kelotok rounding the bend came long after the sound of their boombox magnified its tinny rock and roll up and down the peaceful river and in and out of the riverine trees. All of us in the water and on the dock could see them lolling on both decks amid ladies in shorts or tight jeans with enormous amounts of jewelery and make-up. These were hardworking dames who climbed off the decrepit boat in high-heeled slings looking as if safety was the last thing on their minds. Based on my new theory, I decided they were in search of excitement. Money and attention. Then I remembered that they were earning a living and threw them back in the "safety" net. The men, on the other hand, were definitely dressed for the wild side. They wore unbuttoned tropical shirts, white pants, white shoes, and the effects of too much beer. As they disembarked, one of them held the huge boombox clutched to his skinny chest, while the boat driver ran after him, begging him to turn it off, shouting the equivalent in Indonesian of "Against the rules!"

The tour group staying at the Rimba, including the boys who'd played Ping-Pong with Kate and Amy, had also come up to Camp Leakey. Most of them and most of us and all of the boats' crews were in the water, swimming just off the dock. Uranus loomed overhead at the top of the old ranger tower. He's fourteen years old, the son of Unyuk, and perhaps he'd never seen so many *Homo sapiens* in the river at one time. But I felt safe. What crocodile would dare to intrude on us?

The effect of the Koreans and their hired women was far more upsetting than the thought of crocodiles. The boat stank, the music stank, the mood of the group was all wrong. They had arrived in time to see the afternoon feeding, and the rest of us were determined not to join them. I did not even like the thought of the orangutans meeting these representative *Homo sapiens*. When they had disappeared in the trees, most of us slipped back into the water. My territoriality made me sympathize with Biruté. She had been here when there was no dock, no outsiders, no radio, no clock. Seeing Camp Leakey with all of us scattered along its flank would be like seeing a daughter seduced by a hockey team.

Later, while we were sitting in the dining hall, I looked through the guest book. I had signed on line 490 six months before, and since then there had been 209 visitors. It was strange to see all the names of the summer before. Giovanna and Louise and Eric and everyone else. It was strange to think we had all sat in this dining hall together and impossible not to think that there was some imprint, some presence, left in the camp and forest by our journey through it then. Is it possible to touch a place and leave it unmarked? We must leave cells, thoughts, moods behind. We must leave shapeless pieces of frustration and joy and ambivalence.

"I hear you had a little trouble with Siswi this morning," Mr. Ralph chided, when we had all gathered on the dock late in the afternoon. Kate and Amy had brought their laundry and we'd settled down to scrub. I'd been glad that he hadn't witnessed my chagrin, that he'd been off changing lenses someplace. It's hard not to indulge in one-upmanship in the rainforest. I knew Mr. Ralph would never put a camera or notebook or anything else on the ground near an orangutan, but as I looked at him, standing tired and hot on the dock, I didn't feel as humbled as I might have. The first thing Mr. Ralph learned to say in Indonesian was "I can't swim," yet these ablutions were what made the blistering walks through the forest bearable. I couldn't imagine being too dignified to stick my feet, at the very least, in the Sekonyer.

Then I remembered that Mr. Ralph had learned to ignore the weather, and that in spite of the likelihood of drowning, he had traveled this river and plenty of others on more leaky sampans than I would ever see in my lifetime. So, with a little show of humility, I asked him if he'd like a painting by Siswi. But he said he had plenty of art by orangutans, tipped his baseball cap, and smiled no thanks.

Loyalty

We had no choice but to follow the Korindo businessmen downriver, watching them dance on the upper deck, their silhouettes jerking spasmodically against the sky while empty beer cans flew off to starboard and port, leaving a crooked wake. Then, at the end of the day, we heard a soft purr, and shielding our eyes, we stared and squinted and brought into focus, through all the overhang of sinking sun and pandanus, the prow of the shiny *Garuda II*. At last! I happened to be sitting on the prow of our kelotok, with my feet dangling above the water, drinking one of the beers Mr. Ralph had brought on board as his contribution to our trip. When I saw Biruté's kelotok, I panicked. This was not the way I wanted her to see me! Shoving the beer at Riska, who never drinks a drop of the stuff, I combed the hair off my face with my fingers and tried for a look of scientific detachment. The look of an old hand. Directly ahead was the woman I'd been following for over a year. I could see her on the upper deck. There were people draped around her — children and adults — and they had their own big boombox clanging loudly. Although the park was closed now, and it was against the rules to enter it, the

Garuda II was charging triumphantly upriver with thirteen people aboard, all of them bound for an overnight stay at Camp Leakey!

It's an impressive thing to see a large boat grind to a halt and turn completely around, but Biruté must have given her driver the sign to come about, and there, mid-river, we also slowed and somehow came to a halt. When the two *Garudas* were nearly abreast, Biruté glanced briefly at me, leaned over and spoke without greeting or smile, looking straight at Mr. Ralph. What she said was, "Are you coming?"

Mr. Ralph, who had just left the park after waiting for her for three days, and who was seventy-five years old, put his beer down and said, "I'm . . . I just . . ."

"Oh, whatever you want," Biruté said irritably. "But you're coming tomorrow. Early."

"Of course."

"And bring those pictures you numbered. You have them, don't you?" For a second, Biruté cast a second look in my direction. "Hello," she muttered, but in another minute she was charging upriver again. The movie star sitting next to her was Stephanie Powers.

"I thought you said you'd met her," Kate said, sounding disappointed. I had told her about my book, the dinner in L.A., my hopes of another interview.

At the Rimba, Mr. Ralph shouldered his heavy bag and turned. "You going up tomorrow, then?"

I told him I would love to, but I was stuck. The *Garuda III* was going back to Kumai to celebrate the end of Ramadan. So was everyone else. There were no boats for hire.

"But I have to get up there."

Riska pointed to the woebegone Rimba kelotok, an infamous old clunker that was lying on its side in the weeds. "You could try to rent that," she suggested.

I said, "Right. You're a guest here. Go for it, Mr. Ralph. And if you get it, come down to Tanjung Harapan and pick us up. We'll go with you. We'll be your assistants."

That night I had a wonderful dream. I was walking around a glamorous hotel in a sarong tied at the waist, bare-breasted. The hotel was an assortment of walls that looked like hallways, hallways that looked like walls. I looked into a mirror that made strangers appear to be my friends. Then I could see the faces in the mirror pulling off thin rubber masks. There is a theory, put forth by Anthony Stevens in his book *Private Myths*, that dreaming is part of our evolutionary process, that we *Homo sapiens* have particular neuronal configurations that record and classify sensory information, turning color, smell, and movement into perceptual maps. "Only in humans," he says, "and to some extent in other apes, does higher-order consciousness emerge. It depends on the evolutionary development of language and symbolization, the capacities for generalization, reflection, and self-consciousness, all of which are associated with the development of the left cerebral hemisphere in man." At some point in ancestral time, we primates joined perceptions of the present to memories and dreams and became *conscious* of our thoughts.

Dream: Life forms six hundred million years ago, and the life span of a species is from one to ten million years. There are five episodes of mass extinction. The sixth, ours, includes massive destruction of rainforest habitats where most of the earth's biodiversity lives. The time to repair loss in diversity after each episode of destruction is five to ten million years.

The story of the ark is a fable for our time, the dream of all people, who have similar stories everywhere. Stories of saving may even be part of our biological inheritance. While we sleep, the brain cells that hold our maps are working, communicating with each other about meaning in order to advance our hopes of survival, carving ever more complex maps in our brains, carving frightening and protective thoughts, as one who carves a crocodile and lets loose its spirit.

In the hope that I would make contact with Birutė, that she would finally peel off her mask (and reveal her spirit), Riska and I actually tried to become Mr. Ralph's assistants for a day. When we arrived at Camp Leakey, after a ride on his Rimba kelotok (and still in the company of our new friends Kate and Amy), he carried a small overnight bag that he brandished as he led us up the path. We were lugging all his heavy equipment. "I'm a card-carrying member of the OFI," I assured him, since he seemed reluctant to lead me anywhere near the sacred team house where Earthwatchers and OFIers used to stay. Maybe he was a bit let down, having been informed by a young Englishwoman on the dock that Biruté wouldn't be needing him until three o'clock, which was many hours away, but he marched on along the path past the dining hall into the trees. Biruté must have been somewhere nearby, in the team house or even in her own reopened house, but Mr. Ralph motioned us onward, up the next path, when just at that moment we heard Pak Akhyar call out. He was telling Mr. Ralph, who speaks Indonesian, not to proceed.

This time a shadow crossed Mr. Ralph's lined face. "Let's humor him," he said, turning around. "He doesn't know the Professor invited me. He'll find out."

We all stood awkwardly under the dusty trees. Riska and I were still loaded with photographic equipment, and it seemed we would have to stay loaded indefinitely. Then Biruté appeared, as if to reinforce Pak Akhyar's warning, walking toward Mr. Ralph. "Leave your things in the dining hall," she said without any salutation. Then to Riska, who had visited her house several times, "Do I know you?"

Not a word to me.

Kate and Amy were witness to all this and I felt

thoroughly ashamed. The great person, so much in my thoughts, had not even deigned to speak to me. I realized that Biruté had not been testing me. It was I who had been testing Biruté. I had wanted her to live up to my idea of greatness, nothing short of that. My embarrassment was for her.

Mr. Ralph told us to go on, he was going to wait, so it seemed there was nothing to do but leave him to recover his dignity. We decided on a walk with Kate and Amy, who were trying to soothe my own ruffled feelings, as if I had some shred of dignity to recover, too. Trail No. 1 was hot, but we walked for three hours, and there I was again with three young females climbing over slippery logs, plodding through ankle-deep mud, and studying tiny bees in a tree tube full of honey. Four sets of woman legs, four woman voices, and the struggle to look, to continue, to breathe. This resemblance to my earlier trip both saddened and gladdened me. Maybe there will always be young women to adopt. Maybe that is what I will be left. Maybe this is why I came in the first place . . .

At three o'clock, we went up to the feeding, thinking it likely that Biruté would be on the old bridge, since, if she wanted to catch up on the orangutans, that was where they were sure to be. I don't know why I was still pursuing her, but we headed for the feeding and Biruté never came. Tired but alert, when I saw Emmy and Muchtar on the path to the bridge, I carefully backed away. *Never initiate contact . . .*

But Emmy didn't appreciate my concern for the rules. She wanted to be carried, and as she grabbed at my arm, I was visited by that dream of the infant again. *I was holding Biruté's child. And Biruté. Would she recognize me?* Then Emmy tightened her grip and sank her teeth in my thigh. I had on wet shorts (I'd been swimming) covered by a batik dress, but she bit through both, broke the skin, left a set of toothmarks and terrific pain, then dashed off to the crowded bridge. Trying not to cry, I limped along, too, and saw that Emmy had already found a small can and was dipping it into the milk pail and tipping it up to her lips very delicately, as if determined to demonstrate the social skills she had learned from the *Homo sapiens* who adopted her.

Barn Light and Betel Nut

A stranger or visitor might, however, load a diary with anecdotes of Dyaks who going to the woods, becomeing orangutans, and after several years, having borne many children, have returned and reverted to their former conditions.

— James Brooke

"Never mind," Riska had said, as we passed the Rimba several nights later and saw Biruté sitting on her private deck with Stephanie Powers, sipping drinks by candlelight. "We're going to a better place."

We had another week together before I would get back on a plane, and we were going to a Dayak village.

But first, it is necessary to imagine this village on a riverbank without a road or a car or even a machine, with nothing, in fact, to perform its labor except fire and the strength of hands. The only motor is an old, moody outboard. The only light is provided by sun and kerosene. Everything that happens to a person happens to the place. Birth, death, and everything in between are events that concern every member of the village. One afternoon a Dayak comes out of the forest, returning home after months of cutting down trees. He is tall, tired, lean. His teeth are red — betel stained.

If it seems wrong to describe a place according to what it isn't or doesn't or hasn't, I excuse myself by saying that as an outsider, I was forcefully struck by my sudden release from conveniences. I was not the first outsider in the village, but I was one of the first, and I was used to a world where, even if I ate lunch on a log by a river, I was only minutes away from a telephone. In the village of isn'ts, with only water and trees, I joined a gathering of humans who had struggled to bring what is there into existence. For them there is no sense of *lack,* but rather an enormous sense of regard for their mutual efforts. There is no newspaper, no store. What brings news and nourishment is entirely alive. Roosters. Birds. Trees full of fruit, padis full of rice. And doors thrown open to the light.

"I knew you were coming," said the *tuan* of the house in which we were to live. "I saw a bird flying off the roof this morning . . ."

The plan had been to visit Kudangan, Riska's village, but there was not enough time. At this season, the one logging road would be too wet to be driven, and it would take too many days to travel by boat. So we had opted for Bakonsu instead, a village that has no access by road. And with only six days of my trip left, we had to go by speedboat or not go at all. The river we rode is called the Arut. Then it is called the Lamandau, which breaks into countless branches, each with a different name. It has narrow passages overhung with thick, head-cracking green, and we saw signs of lumbering all the way up and all the way back. The biggest cleared places — brutally struck by Korindo — looked like war zones.

In the speedboat, we were soaked by rain. Passing through the old Kotawaringan kingdom, we stopped at a place where it was possible to buy bowls of hot soup and a bottle of gasoline. Occasionally the river widened. Occasionally there was a barren lumber camp. The buildings in these camps are often hallowed, prior to erection, by the burial of a head, although headhunting is strictly illegal now. Nevertheless, without a head given to the ground, workers refuse to work, so it is said. The site would be too unlucky. Because of this, Dayak children are not safe, for they are the most accessible victims. Or are these apocryphal tales?

The second time we climbed out of the speedboat, thoroughly soaked by rain, it was through an old kelotok that was tied to a dock. This inelegant arrival was witnessed by two serious children who stood waiting at the top of the bank as they had been doing every day for three or four weeks, when Riska had sent a message upriver that we might be coming. Now, after all the anticipation, the children — ten-year-old Yefni and her seven-year-old brother — were calm, coming down to take our packs and help us up to the longhouse where we would be staying.

It was not Riska's village, but her grandfather had been known here, for he had built the schoolhouse and the church. The house where we stayed is one of the oldest in the village — perhaps a hundred years old — and it's owned by Pak Pangi, an elder who allows his nephew to live there and look after it. This was apparently both an honor and a burden. The house is made of ironwood, which is almost eternal, but the roof was disappearing at a mortal rate. As we struggled up the riverbank and followed the children up a path, I saw first this great bowed roof and then the long, graceful house underneath, wider at the top than at the bottom and braced by its many wooden legs. It was close to the river, to the graves, to the "offering tree."

It was dark enough, and its wood worn soft enough, that it seemed an altogether enchanted place. In the daytime, it was like the inside of a barn, the only light slipping through chinks and broken boards. In the evening, as it got darker outside, Yefni's mother — referred to, in the Dayak way, simply as the mother of her child — lit a kerosene lantern and hung it in the center of the house. Then the shifting floor with its sheets of dark orange and green linoleum held pools of surprise, pools of nowhere, pools of orange tarp and grass mats. At the back of the huge single room, there was a curtain of fabric that could be drawn across a rope to make a little privacy. That was it.

Wood creaked, chattered at night, shone, absorbed rain. Bamboo slats underfoot, not nailed but tied so they moved as we balanced on them, clicked softly as we walked. Walls, missing pieces, offered now and then a slice of outside. In the dawn or before, the rooster beside you makes a triumph of announcement: Now now. There is no noise. Then there is a wakening. Gibbons in the forest, roosters and hens and birds. Somewhere a pig, somewhere a cow. Insects. Women pounding rice.

The kitchen was lower than the rest of the house and was divided from it by two boards at knee height set between the thick poles that support the roof. No nails, remember. No furniture. Not so much as a stick. No decoration except for some pictures and letters painted on the walls. We sat with our backs against these because they sloped out, because the house was widest at the top, because it felt good to the spine. In the kitchen there was a raised hearth full of sand and ash and a fire, a kettle very black and lidless on three stones. There was a dish rack with empty bottles on its pegs. On the floor, catching rain, were many buckets and two antique Chinese jars worth a small fortune anywhere.

We entered the house, took off our sandals, brought in food and packs. I blinked and studied the dim surroundings. No fuss was made. We simply sat. Then, in the daytime darkness, stretched out on the slatted floor, we slept.

Later, Pak Pangi took us into the forest and out to a ladang, which I wouldn't have recognized as anything planted on purpose except that two women were harvesting rice there in hats so wide-brimmed they were like roofs carried on the head. "Rice, rice!" Riska chimed, for more than the forest, she loves a full-grown ladang, and we had come just in time for the harvest. Following Pak Pangi and a herd of helpful children, we stepped over several ominous-looking insects.

"Is that a scorpion there?" I asked once.

"Oh no."

We stepped over several more of whatever they were, with several children crunching along in bare feet. "Is it a centipede?"

"Oh no. It is different. But much more poisonous."

The kids walked quite happily through these much more poisonous whatevers and through fire ants as they picked bright orange mushrooms that we carried along to take back for dinner. Pak Pangi stopped frequently to pick a leaf of one plant or another and describe its usefulness. "We use this for the stomach. This for pains in the joints. This for prostate trouble . . . dysentery . . . incontinence. . . . Because we have no doctor," Pak Pangi pointed out. One of the children following us had open sores all over his head, but the others looked healthy, considering the muddy river in which we had swum in order to wash and cool our sweaty bodies.

We were walking through a forest market where everything is given and where everything has usefulness. There is the material with which to build roofs, floors, docks, walls, ossuaries, graves. There is bamboo to be eaten, to be made into utensils, to be made into containers and floors, to be woven, to be burned, to be sold. There are herbs and fungi, fruits and flowers, tubers, palms. For those who live in this forest, each daily walk breeds familiarity, breeds claim, breeds saplings, as seeds are moved and dropped. Each day the forest is reconstructed in the mind. Mapped and learned. Everything in the forest and mind reacting to one thing: light. Light being sought or avoided. Heat needed in minuscule quantities. Or shunned.

After three and a half intensely hot hours we circled back to the village and stopped at Pak Pangi's house, with greenery on all sides and the river in front. He says he's seventy-four and his wife is seventy. She had a sixteen-year-old daughter leaning against her shoulder and they both laughed at my incredulity. Pak Pangi says his father died at 145. He has thick black hair and youthful features and is obviously pleased with himself. The secret of his youth and vigor, he told us, was over there on the table. With that, he brought a covered glass jar across the room; it was large enough that he had trouble carrying it. The contents included a compote of fetuses and ginseng floating in arak. There was a fetal crocodile with its long, unborn mouth shut tight and a fetal mouse deer and a fetal deer curled inches apart, as well as a horn covered in fur and some leaves of an antimalarial plant. Politely, Pak Pangi offered me a glass, telling me, however, that the biggest secret of youth is the consumption of live mice.

In the middle of the room was a twisted, ropy stick, described as a gift from the gods to human males, but a secret gift. His wife was smiling as if she could vouch for this, but when I asked her if she drank the concoction in the big glass jar, she shook her head emphatically and said she had her own medicine. Then winked.

We had come almost empty-handed because of the limitations of the speedboat, but there had been time, in Pangkalan Bun before we rented the boat, to buy a few gifts and have a few of our pictures developed. The gifts were notebooks and pencils for the children. The pictures we had put into a plastic album. They were mostly of orangutans or of orangutans with us, and these latter afforded the villagers much pleasure. Invariably, as they came to the house, climbed the first notched log to the first porch and then the second notched log to the doorway, announced themselves, came in and sat down with their backs against the wall, the little album was brought out. And invariably they pointed at me and at the orangutans and said hysterically, *"Sama, sama!"* Same, same! At that everyone would break up. I, too, would laugh, but in some bewilderment. (I hoped the similarity was because of my reddish brown hair, and not for lack of a visible neck.) Laughter, like tears, is infectious, and in this longhouse everything is shared. Yefni takes a small piece of watermelon from Riska and cuts a piece for the baby, then the rest in equal shares for her brother and herself. She gives most of her own piece to her mother. If a visitor brings the children a treat, they touch everyone in the room in order to spread the pleasure. There are no toys in this village and there is no dearth of happy children. They are included in everything. They take part in work and conversations, the two occupations of everyone who is not asleep.

Sleeping goes on any time, any place. One simply stretches out. One simply goes to sleep. There is the floor, where one sits or lies, and the ground, where one walks once one has reached the proper age. Triana, the thirteen-month-old baby, was almost never put down. Her mother, who pounds the rice outside in the morning before anyone else is awake, might lie down with her to breast-feed or talk or sing or sleep. Most of this mother's day passes in visiting with people who come by, and playing with her children (there are three). I never saw her eat.

Because it was harvest time, all the women were pounding rice while I was in the village. Rice lay drying on mats in front of every house, waiting to be threshed, then it was carefully poured into storage baskets and taken to the rice barns, which were small versions of houses on the wooden stilts that keep out mice and water.

Yefni's mother and the others were giddy with the pounding, the sounds of which, rhythmic as heartbeats, thump-thump, thump-thump, went on somewhere in the village all night.

Soon after we arrived there was to be an engagement party, and the elders decided to provide a welcoming ceremony for us at the same time. This would include a hornbill dance, and it was suggested, during an official visit by one of the elders, that I might contribute to the funds for rice wine. When enough money had been collected from various families, the elders went from house to house and bargained for the home-brewed wine that is kept in antique jars.

The idea of a party was exciting, and Yefni went through her clothes, which were stashed in a basket. She takes care of herself — does her own laundry, combs and delouses her hair. On the afternoon of the party, it rained heavily and the buckets and jars filled up. I looked forward to a rainwater bath and took a small pail of the precious stuff into the falling-down room that's attached to the kitchen where the rooster and hens sleep at night. Here, with all eyes averted, since there is no real wall, I took almost everything off. It isn't done this way, but I did it.

Everyone in the village bathes in the river twice a day, the women washing children and clothes as well as themselves. The way it is done is to soak in your clothes (at least your sarong), then take them off in the little hut on the dock and change into dry ones. In this village there are many frail docks attached to the steep riverbank, and on each dock one of these huts where it is possible to change clothes or to squat in privacy over a hole that goes straight to the river. When I heard he had traveled for seventeen years among the Dayak, I had asked Mr. Ralph about the problem of shitting. "Where do you do it?" I asked. He told me to find a tree, and added, "There is always a pig nearby to clean it up." I don't know whether there was mercury in this river, or even crocodiles. It was full of mud from the runoff of timberland; I didn't notice anything else. There were a few small kelotoks every once in a while, and a speedboat much less frequently than that.

Riska and I took Yefni to the celebration, since her mother doesn't leave the house untended. She was born in another village, and maybe for this reason there has been a long-standing feud with one of the neighbours, during which the toes of her hens were once cut off. (Among primates, it's usually the females who leave the natal place to mate, and this is nearly always a disadvantage.) Since her husband had been in the forest cutting trees for several months, she hadn't even visited her own ladang across the river to harvest her rice. So she was firm in her refusal to go with us to the party.

Once there, we climbed the log ladder, left our sandals outside on the porch, climbed the next log, and entered politely, heads down. The house was full. An entire gamelan group was clustered at the entrance along with all the adolescent males. Older men sat with their backs against the brace that runs down the middle of the house at floor level. Young women sat along one wall, busy with children; older women sat near the back of the house, busy, like the men, with betel-nut containers. Each conversation or greeting, however small, was preceded by the ritual Dayak greeting in which the hands are brought together, then raised to the sides of the face, after which the right hand touches the heart. This genuflection gives an air of sobriety and reverence to even the most drunken encounters, many of which were to come, but in the beginning it was a rather quiet gathering. There was anticipation in the house — especially as the first notes were struck among the gongs — then a rising sense of excitement.

There were also going to be plenty of speeches, during which Riska leaned against me and whispered her translation and Yefni clutched my hand, but only after we had all assembled ourselves. Then the elders described the soon-to-be bride and groom and offered advice to each of them on their engagement. There were admonitions to the rest of the village, especially their friends. By virtue of this ceremony — celebrated with Kaharingan mantras as well as Christian prayers and hymns — the couple would be set apart. They were not to be flirted with by any of the others, nor were they to take any sexual liberties themselves. (Penalties for infractions involve the payment of antique jars.) A small table held a candle, a crucifix, and a sheaf of rice in a glass. Also a prayer book and two bowls, one for gifts of money from guests and one holding two gold rings. All this was as nothing compared with our welcoming ceremony, however, which involved each important man of the village getting to his feet, introducing himself, and then explaining his position, age, and family to me. I recognized the chief, to whom I'd been formally introduced, and Pak Pangi, of course, but Dayak villages have various civic functionaries. The chief I had met was appointed by the government, which gave him only one kind of authority. Another chief is appointed by the community. I began to lose count as a large bowl of cloudy rice wine was offered, and whispering, Riska admonished me to "Take the bowl, yes. Now drink it; it's delicious. Right. Now hand it back." Happily, our relationship had changed. The quest was in some way mutual now. I had a lifelong desire to experience life as it must once have been lived by human beings in every place — the primal experience: a settlement, a tribe. I imagined, as Louis Leakey once had, that our origins were to be found in the shade of a great, vanishing forest. Riska may have represented some of that to me, but there was no simple conclusion to be drawn from seeing her in situ, except that she is already nostalgic for many of the same things.

After all the men of distinction had spoken, it was my turn. I stood up, towering over the assembly, and said that I'd come a long way by airplane because I had always hoped to meet the "true people" of Kalimantan (although I might have said the "true people" of anywhere). And because I knew that it was especially surprising that I was traveling alone, I mentioned my children, hoping in that way to connect myself to the villagers, because I felt slightly desperate. Massed around me were a hundred or more human beings who found me as peculiar as a Martian and who expected profound revelations, at the very least. Among the Dayak, young men acquire prestige by traveling. In former times, since a boy would never be accepted as a husband without taking at least one head in a raid or ambush, trips were often made for that purpose. The severed heads were essential to the health and prosperity of a village. Another reason for travel was the acquisition of goods. And experience. Upland men have always traveled toward the coasts in search of manhood, but women didn't have such quests. Do they anyplace? To my hosts, I was an anomaly and probably suspect. Anyway I was far, far from my own range.

I thanked everyone for the kindness they had offered, saying that I had found not only the true people of Kalimantan but the most hospitable people in the world. When I lowered myself to the floor, I did it gingerly, pulling my skirt down over my knees and trying to decide whether, in the midst of the boys in their leather jackets and jeans, I had really come, at last, to a place where people still live in the ancient way.

Passed from one generation to the next, this culture had pottery and domesticated pigs, cloth made of bark, rattan and pandanus weaving. The Austronesians had cultivated tubers and palms and practised burial in urns. The holy buffalo had traveled with them, and the spirits they brought settled into the trees and rivers so the ancestors of the ancestors became deities, and from this time and forever the dead have been buried with gifts. From this time and forever life spirit has been present in severed heads.

From the Dong-son of Indochina came the bird, the reptile, the bamboo. From India came the serpent of the underworld, the hornbill of the upperworld, the tree of in-between. As settlements grew, they began to forge metal tools, and with them they extracted starch from the sago palm and carved the most efficient blowguns in the world. With the tools they began to cut down trees and burn them to fertilize the soil. They planted rice. With the burning of trees and the taking of heads, the rice grew. Then the number of settlements.

During my stay in the village of isn'ts, I knew who I was and I knew that I belonged, at least for the time I was there. I was taken on faith, which is quite an experience. I had made an exceedingly long and arduous journey, and they were honored by my presence. They were going to considerable trouble to welcome me, and I was honored by their efforts. We were held in a web of mutual esteem, and all of it was serious because in a village on the edge of a wide brown river everything we do and perhaps everything we think affects everyone else there forever.

Throughout the party, I asked Yefni to stay next to me so I could hold her hand. I suppose this was because all the women and many of the men were holding children, and I felt naked and unsexed. Yefni was very quiet and serious, even when all the other children began to sleep, as if she had the responsibilities of a caretaker. The speeches had followed a meal of pork and bananas and rice placed bowl by bowl along the floor by the younger women, while the older ones still sat together chewing betel nut and enjoying themselves. The rice wine was slightly sweet. It was delicious, but I was aware of myself as a representative, though I wasn't sure of what. Perhaps the nicest thing one can do in such a circumstance is to make an ass of oneself, to kill any romance with the outside, to warn these good people away from it. Nevertheless, I kept my knees together and my legs well tucked until I was led over to a large bronze gong and told to sit on it.

With Riska perched on the other side, we sat back to back and managed to cling, more or less. Behind us, a huge Chinese jar was draped in yellow cloth. We were draped in it, too, and a bearded man sat in front of us, holding a rooster by its legs. On a table there were two bowls of rice with a wide-bladed knife lying in a metal ring. The situation of the rooster looked ominous. Someone was chanting. There were three men crouched in front of us. On the rice in each bowl lay a thin pandanus bracelet with a tiny leaf packet of rice attached. Once tied on, the bracelets were to be worn for three days or this welcome would be forfeited and certain penalties would be charged against us.

The knife was suspended over the rooster. The chanting was furious, and in the background there was now the astonishing clamor of gongs and drums. The beat was so intense that our bodies were nothing but skins stretched over our bones, drums upon which to feel the rhythms reverberate. It made for a deafening racket, and there was still the rice wine being passed, which was going to my head, along with the taste and feel of betel nut. Balanced as I was on the ancient bronze gong, I realized I had never been part of a real ceremony in my life except for my baptism, which I couldn't remember, and an unlikely wedding in a Tucson hotel. Never before had I been touched by the sword, and even as I was overwhelmed by the weight of this realization, the rooster submitted to the knife; his comb was cut, and the bloodied knife was placed against my tongue. Gongs rang out and glasses of wine were raised and the bracelets were tied around our wrists.

Our welcome was followed by the hornbil

dance, which is the touchstone of Dayak social communion. Here, the arms are raised slightly behind the back and the legs are bent to make the body glide and the whole business is hierarchical and delicate. I was offered the first scarf and danced with the chief, his wife, and one of the older men. As the chief's wife affected a look of stern nonchalance, I attempted the same look while raising my arms behind me like droopy wings and lifting my feet in a more or less birdlike dance. The chief's wife moved so slowly that her glide was almost imperceptible, while the two men swooped between us, and one of the teenagers, a boy in a black leather jacket and bare chest, swept in and out, offering each of us long drinks of wine from a glass. It was incumbent on us to accept and to offer back.

While the boy in black leather swept in and out, beating his feet insistently on a thousand loose bamboo floorboards, the other boys were doing the same thing at the front end of the longhouse. They smoked and stomped and swayed in their bare backs and stomachs, and it was only surprising that the girls were able to sit so disinterestedly at their feet, although the mothers at the other end of the room watched their sons with passionate interest. As each quartet of hornbills finished, the scarves and sarongs were formally passed to others — the gestures of hands against face, the slight bow — until more and more people joined the dance and it became, with the racket of bamboo and the onslaught of wine, wilder, wilder.

Around midnight we climbed down the house poles in the dark and made our way home. Yefni was asleep on my back, my arms curled around her knees like protective wings as we felt our way step by step in the dark, which reminded me of another walk I had made, in Bali. In front of the house I stopped and looked up at a perfectly full moon perfectly placed over the roofline. In front of the moon stood a pointed tree, black, as if cut out of carbon and pasted to the pale sky. "Look," I whispered reverently. "Look how beautiful."

Dropping off my back, Yefni opened her eyes. "Oh, it's nothing," she said. "Just the moon. Come inside."

Because longhouses are entered via two poles, one to the porch and one up another level to the house itself, the first sight of a visitor is the top of a head in the doorway. Then there will be a voice, maybe a tap on the wall, and often a gift will be laid on the mat or the floor. Our doorway had a gate fitted across it in order to keep the baby, Triana, from falling out. The long boards at knee height between the kitchen and the rest of the house were there for the same reason, since the kitchen was lower and Triana might have fallen into it. These houses are built with babies and animals in mind and also as protection against headhunters, who made raids until very recently.

Maybe headhunting and gift giving spring from the same source, now that I think of it. For thousands of years, headhunting was part of Dayak life — a system of replenishing the energy of a family or village on the death of one of its members. In this village, they say that headhunters are men and boys who go a little mad. They're bitten by an addiction to certain substances and can no longer be trusted, even by their own mothers, they are so hungry to take heads. Our ancestors all over the earth believed that the head and spine — the marrow of a man — were connected to his essence. The Polynesians called it *mana*. In Dayak villages, there was usually a head house for these powerful relics, or they might be kept in the rafters of a longhouse. (Chanting mantras, Pak Pangi had climbed a wooden pole in front of our house one afternoon and lifted the only head in the village from its resting place on top of a jar in a crumbling ossuary. "Taken in 1500," he told Riska, handing the skull to me.)

The visitors to our house did not come to take my head, but they certainly wanted to have a look at it. Along with curiosity, they brought food. A widow had spent all day weaving two baskets and had filled a third with her own rice. One visitor — Pak Pangi's son — brought a basket his mother had made. It was covered with shells and made of beaten bark, and I coveted it. But it was not a present. He was offering it for sale. This is the only time anyone in the village tried to sell me anything. I got the sense that money was not much valued. More surprising, objects were not displayed, nor were they valued for beauty. What counted was the history of a thing and its line of inheritance.

In our world a present is selected to match the recipient's desire or to express the giver's taste. In either case, it is, as we say, "the thought that counts." Issues of value come into it, and issues of reciprocity, but rarely do we give because it is an honor to pass the heritage of our family into the hands and home of someone else.

In the longhouse, it was good form to reciprocate when a gift was received. To the widow I gave two packages of ointment. Riska had insisted that we buy several of these because, she said, people in the village like to have something to rub on their chests when they have a cold. I began to regret that we hadn't brought other things, especially medicines. This was the only thing anyone ever asked of us. One afternoon as we were walking through the village, a woman approached us with her hands on her belly. When she saw us, she lifted them to her face and greeted us, then told us she was on the way to the clinic; she was in pain, but she knew they wouldn't give her any good medicine; they never did. Then, we passed her on our way home, when her hands were again on her belly and she looked at us miserably. "Do you have any good medicine for the stomach?"

"Didn't they give you anything?"

She held up a packet of six or seven capsules.

"It's just tetracycline," Riska said in the same tone Yefni had used about the moon. "That's all they ever give."

The only medication I had with me was the much-squeezed tube of first-aid cream I had been looking for in the dark the night I rubbed toothpaste all over my backside. I'm not sure what the cream was actually for, or why I had packed it, but I gave it to Yefni's mother and she was glad to have it, saying she would share it with anyone who needed some medicine. I kept thinking what a huge difference I could make to the lives here, although surely the same opportunities for generosity exist on the streets around my house in Toronto. There is, however, a glaring truth about gifts. Toys bought in Pangkalan Bun or anywhere else for Yefni and her brother would not have brought them happiness. Toys would have separated them from their friends and even from their elders, who regarded them as important members of the village, listened to them, talked to them, held them and walked with them, shared work with them, and depended on them. The children, it seemed to me, required no gifts. Nor is this a place where adults long for ornaments. Yefni's mother seemed very happy to have my batik dress when I left, but she was careful to distribute other things we brought among friends and neighbors. The notebooks and pencils were given to the school to ensure that her own children's friends would not be particularly favored.

At night, she lit a faulty lantern after a struggle of pumping and blowing. In the only complaint I heard her make, she admitted that her husband had traded their good lantern and come home with this one, and the tone of the admission was both frustrated and indifferent. I was doing so well with tone that I was actually glad not to have any shared spoken language. After a few days of listening, I found all the nuance of language had begun to make sense. I was drifting backward. I was becoming prehominid, watching a woman wipe the glass of the lantern, push and pull at the pump, trim the wick and then jump back at the tiny explosion. We sat or lay on the floor and engaged in life minute by minute.

Despite the intense heat of the day, during which we lay on the floor every afternoon in a sweaty torpor, by midnight it was chilly. We went to sleep early, by eight or nine, and when I awoke during the night I pulled whatever I had in the way of clothes out of my pack and put them in a pile on top of me. (The fleece jacket had been abandoned in Pangkalan Bun and even now would have been ridiculous.) Growing warmer, I lay on the floor under the steep roof and listened to Riska and Yefni's mother talk. It brought me great pleasure to hear Riska using her own language, and I wondered whether what she had told me in English about her life had the same flavor when she told it in this setting.

One morning, since she was longing to get to

her ladang, I suggested to Yefni's mother that we help her harvest her rice. It made her nervous to leave the house unprotected, but the temptation of assistance overcame her, and eventually we locked the door and went across the river in a borrowed canoe. There were two or three ladangs within shouting distance, and each had a small, roofed platform, or *pondok,* on wooden stilts. These were used for sleeping or resting in the shade and for storing the enormous hats we took from hooks and wore into the field, walking single file, heat hitting us like stones from above and rice singing its dry song. There are years when an entire ladang is overtaken by worms or when rain breaks the stalks or when someone's cow wanders into the field and crushes them flat. One family is wiped out and a few feet away another family thrives.

The tool for rice cutting is half-moon-shaped and cupped in the palm, and slightly lethal, so I opted for Riska's Swiss Army knife and tossed the rice heads into a large back basket planted on the ground. Rice was growing haphazardly, as if planted by the wind instead of human hands, and everyone in the fields was excited to see big, tall me out there in the hat cutting rice with the best of them.

I was sweating but not hot because of the hat, and when we had worked for a couple of hours we rested in the shade of the family *pondok,* then picked some wild mangoes by the river and came home with a full basket of sheaves. I was bruised, bleeding, and pleased with myself. Instead of sitting on the sidelines, I had done something useful for a change.

It was after this outing that we were visited by a woman who introduced herself as Pak Pangi's older sister. I'd been ruminating over Pak Pangi's youth ever since we'd arrived, and now I was determined to get to the bottom of the age mystery. So when the old lady came up the pole and sat down on the bamboo slats, in spite of my strong desire to nap I asked Riska to ask her age.

Riska leaned forward, introduced the topic politely, and turned back to me. "She says sixty or seventy," she reported with equanimity.

"But Pak Pangi says he's seventy-four!"

Riska leaned forward again. "She says, 'Oh, then he must be right!' "

"How can she be sixty or seventy if she's older than Pak Pangi?"

Riska turned to the old lady and translated my second question as politely as she could. Then she translated the old lady's response. "She says then she must be eighty."

Yefni's mother had been listening to my queries and showed no surprise at the old lady's answers. She got the lantern from its hook and began her nightly argument with it. Evidently it was perfectly satisfactory to have a variability of twenty years in one's reported age. There are no records in these villages, where seasons are more important than years. It was harvest time, not 1996, and what was the matter with me anyway?

Yefni's mother says she's thirty-six. She married when she was twenty-three, and I feel like an idiot wondering what she was doing all that time. I couldn't stay at home past the age of eighteen! I was dying to get on with my big, grown-up life and ran away from college to elope when I was twenty. I'd always thought "places like this" bred early marriages and teenage pregnancy. While I was musing, Yefni's brother, Elias, came over to touch fingers. This was because his grandmother had come in with a treat. The grandmother had also come with another grandchild who was almost Triana's size. After spending three weeks with young orangutans, I sat back against the sloping wall to watch the babies play.

Triana stays close to her mother twenty-four hours a day. She nurses whenever she wants, but is completely "toilet trained" (a ridiculous phrase in the village of isn'ts). Riska was dumbfounded when I asked why the child wasn't wearing diapers, but was either naked or in little underpants. "Babies are trained here by four months!" she protested, and it was true. There was never pee on the floor or anywhere else. Somehow, without words, Triana let her mother know what she needed, then a floorboard was moved a few inches and her mother would sit her down over the opening.

At thirteen months, Triana doesn't talk. She walks, but without much agility, and although she never cried at the sight of anyone else, she screamed if I turned my head in her direction, so that I had to hide my face to make her stop. The dog, too, growled, even when I offered him food, but he followed the children, who kick him and chase him out of the house. Never mind. I'm content to lie on this mat on this floor in this house for the rest of my life. I've brought not a single book to read, I don't even want to talk, and I can't imagine why I'd want to buy anything.

What I'm doing is noticing. It's a full-time job.

And missing my own children. Kristin, especially, would have loved this place. I missed her delight and curiosity and wished that I could magically transport her. Even for a day.

Then *one afternoon a Dayak comes out of the forest, returning home after months of cutting down trees,* and his wife, who has been looking for him for many days, begins to move differently, to talk differently, to smile and show her pleasure in his nearness. *He is tall, tired, lean. His teeth are red-betel stained,* but I am mainly surprised by them as a couple, by the dignity of their happiness and by their suppressed desire. After eight months, the *tuan* has come home to find two strangers in his house, but he has never seen a white person before — "only once, a Dutchman going by in a boat when I was a boy" — and he appears to feel not intruded upon but honored. On the other two occasions when outsiders came to this village, he was away in the forest.

"How much will he earn," I asked Riska, "for the trees he has cut?"

After consulting him, she told me, “He went with a small group. First they have to rent saws and equipment. They build a shelter in the forest. They carry in rice. When the trees are cut, it’s terrible, very dangerous. Then they drag them a long way to the river. All this with their arms and backs. Then they tie them up and float them to a place where they can sell them. Then they pay their expenses and share what’s left.”

“Yes, but how much?”

“He said about eight hundred dollars, perhaps.”

At last the kerosene lamp could be attended to. Yefni’s mother brought it across the room to her husband with a very wifely grimace. He smiled and worked on the wick. Trimmed it evenly. Fiddled with the lantern, laughed at himself, asked Yefni to pick him some betel leaves, chewed, pumped, trimmed, blew, laughed. And talked without cease. When the lantern was lit and casting its light in the gloom, we sat with our backs against the walls and listened to the words. He told us about his father, about ghosts, about headhunters, about many things. Yefni’s mother nursed Triana, and Triana looked at me and screamed.

Yefni herself was told to go to the corner where the clothes were kept and to bring something out for us to see. When she unfolded it from her hand, it was a small Ming jar, very beautiful in its shape and simplicity, like a stone that had been rounded by some river and then hidden away. Yefni's father said that once he had owned five of these tiny jars, that they were his inheritance, but he had given four of the jars away to special friends. "We call it," he said, pointing to the crackled glaze, "a thousand pieces."

"He wants you to take it," Riska said.

It was then — at that moment — that I realized what two trips to Borneo had done to me, for I saw that whether or not the snake eats the frog, they are turned of the same clay. That like my brother and my father I am mortal and no longer afraid. All my life I've been physically awkward, afraid of losing my balance, slipping, being swept into the riptide or off a pair of skis, in short, of being unable to handle myself in the world of water and rocks and ice and trees. But now, older and stiffer, I've lost my fear of being hurt, stung, bitten, broken, or sick. I sleep on a floor as easily as in a bed, and at night my dream soul travels without the heavy baggage of guilt. True, my day soul is still attached. To comparisons, to judgments, to objects. Even to grief. I looked at the jar. I said that on no account could I take it.

Riska said, "He knows he can sell it for five hundred dollars. But it will never be sold for money. He has saved it. To be a gift."

I had been thinking of a way to tell the rest of this story and then end it. How far does this journey go, I'd been wondering? How large is my map? While I was visiting the tiny gold field of

Jemantan on the Sekonyer River, other Canadians were poking holes in the soil farther east. When Bre-X announced that it had found the world's biggest gold deposit in East Kalimantan, and when, later, that announcement was proved false, no one remembered that the River of Precious Stones has no mouth, no source.

Once I had bought a ceramic elephant made in China in a store called Pic & Save. As I had predicted, it fell out of my hands and broke when I parked my rented car, but I'd carried the pieces back to Canada. Now I put a tiny Ming jar in my pack as carefully as I could, wrapped in a sock. The name of its glaze is "a thousand pieces," but I would carry it home intact.

Jakarta

There are many universities in these ape lands,
and at all of them humans are bred and studied.
— Raymond Corbey

"Did you know it starts with a *D?*" my mother asks excitedly, when I call her from my air-conditioned hotel room. "I've finally located you on a map, but I never did find Pangkalan Bun."

"Never mind," I tell her, because at this point I am unaccountably and shamefully glad to be away from that place. For almost a day, I don't care if I never go back. "Don't bother finding it on your map," I say, because I've lived 99.9 percent of my life with running water and flush toilets and these, whether in Jakarta or Djakarta, are suddenly more important than trees. The small hotel restaurant sports a real rose on each table, and after my month in the forest, the one in front of me looks miraculous — the first flower I have seen in weeks. I spend a day in bed watching Celine Dion win a Grammy and reading a fat John Grisham, then I am ready for a return to my life. But there are two things I want to do in Jakarta before I leave. First, I want to see Amy and Kate. It feels good to have people I can look up, with whom I can exchange news and share a few stories. I'm suddenly lonely — a wave of desperation hit me in the Semarang airport, with its memories of Esta and Kristin and our long wait there in front of six enormous air conditioners evenly placed around the glass walls and going full blast. I had left Riska in Pangkalan Bun like another daughter. Already, I wanted some company.

The second thing I wanted was to meet Mr. Suprapto, who was head of the park when Biruté lost her licence. But when I had called him from Pangkalan Bun, Mr. Suprapto insisted that I first meet his boss, Mr. Sumarsono, at the big, central PHPA. Not just meet him, but obtain his official permission for our interview. So I called Kate and, killing two birds with one stone, asked her to help "with Mr. Suprapto. If I can find his boss and get some form of approval from him, will you translate?"

"Sure. Did you ever get to talk to Biruté?"

"No. I only saw her once more, from a distance, sitting on her balcony!"

Friday was spent among glass towers. Inside one of them, shiny with shellacked hardwood of various colors, are the offices of the Ministry of Forestry. Here, on the top floor, sit the boys who hand out timber concessions — the boys who sell the trees everyone in the offices below are working to save. Two hundred and seventy-eight logging companies have concessions to cut down Kalimantan's forest, and they have already cut down almost half of it. (To make things worse, President Suharto issued a presidential decree in 1995 converting 14,000 square kilometers of Central Kalimantan into agricultural land with the plan of moving another 316,000 families into the area from Java.) But like other rainforest stories, this one doesn't appear to be a drama of good guys versus bad guys, since it is commonly understood that the ones in charge of saving the trees are doing their very best to be promoted upstairs.

After leaving a note to Mr. Sumarsono with his secretary, I took an elevator up, deciding to look for the former head of the PHPA, a man named Sutisna. He'd been in charge during Biruté's years at Tanjung Puting, and he'd been Mr. Suprapto's supervisor at that time. Coming out of one of the shiny offices was a man with a prayer mat on his shoulder, who was about to join a group of colleagues by the elevator, waiting with their own mats slung around their necks as casually as locker-room towels. It is every Muslim male's duty to visit the mosque on Fridays. "I'm looking for a Mr. Sutisna," I said.

"He's not far away . . . only five hundred meters, but you would need a taxi to get across all the intersections and past the roundabouts, and anyway he is on vacation. He is working for a company called Pacific Timber, although he is still also on the board here."

Pacific Timber?

Downstairs, I wandered into a fancy gift shop

that caters to the office workers and sells tortoise-shell combs and rings, wondering how many endangered species were ranged on its shelves. But I was in a nasty mood by then, marooned on a marble island in front of which, on a circular drive, were parked the many expensive cars of the well-heeled timbermen. In Long Beach Biruté had told the students that the fate of the wild orangutans is inextricably bound up in the socioeconomic and political forces that shape Indonesia and Malaysia. "Whether they live or die," she said, "will depend on the political, social, and market forces around the globe." Conservationists, she told the students, have to become political. They have to function within the system because "it's human greed which is killing the tropical rainforest and the orangutans in it."

Mr. Sumarsono's office — one of the ones on the lower floors — had a scuffed parquet floor and a leather sofa near the secretary's desk. There was a coffee table with a bunch of artificial flowers in a vase on a lace doily and two metal cupboards with handles that lock. I could imagine why a man working here would like to move up to Forestry, but I decided to sit it out. Sooner or later Mr. Sumarsono would appear; I was sure of it. This tactic caused the secretary, to whom I'd entrusted my note several hours before, to eye me warily. Then suddenly I was granted "five minutes" and hurried down the hall. A door was opened and Mr. Sumarsono — perfectly hospitable and more than pleasant — invited me to sit. He explained that Mr. Suprapto was no longer involved with the park. He said, "It's very difficult for us, working with Dr. Galdikas, but we try." Then he agreed to let Mr. Suprapto speak to me as long as he discussed only the things he himself did while he was there.

On Saturday morning Kate came to the hotel, and we set off together in a taxi. Mr. Suprapto lives a long way from central Jakarta; it took us close to an hour to get out there, with one short stop to pick up a box of cakes. Kate had chosen the taxi carefully. It was air-conditioned and the driver was reasonably sure that he knew where we were going. Jakarta unrolled itself lazily that day — the traffic was bad but not terrible and the newspaper and magazine hawkers were moving up and down the center of the four- and six-lane streets. My young friend had a confidence I can't remember feeling at her age. For the millionth time I wondered what life would have been like if I had lived it differently. Unmarried, working, learning to trust myself.

Mr. Suprapto works in the office of finances at PHPA in Bogor, which made his residence in Jakarta a mystery to me since Bogor is a pleasant town with lots of pretty homes and fresh air and trees. The mystery grew as we passed into and out of the wealthy neighborhoods I had expected of a PHPA man, finally entering an area of small recycling businesses where old bedsprings are turned into something else. How does a man decide to be the straightener of bedsprings? How does he decide to do his work honestly, bucking all chance of promotion and fancy home?

Mr. Suprapto had told Kate to look for a store, and there he was in front of it, waving an arm. It's one of those places that is open to the street all across the front and it's as neat as a pin, full of cakes, candy, bandages, utensils. "I go out on weekends to pick up the inventory," Mr. Suprapto told us, leading us behind a counter to a passage that connects to his house. He and his wife have lived here since he left Tanjung Puting, telling his department that he didn't care what position they gave him because his position in Tanjung Puting was untenable. He had been given a "mission," as he put it, and then been advised by his superior — Mr. Sutisna — to ignore it when it conflicted with Mrs. Biruté. "I really tried to work with her but . . ." Dressed in beige pants and a polo shirt, he looked younger than I had expected, but not young.

When he first went to Tanjung Puting, he said, there were fifty orangutans at Camp Leakey. The official policy was changing. There was concern about disease. Might these ex-captives infect the wild population? Were they more likely to carry disease if they continued contact with human beings? Was the object to rehabilitate them to the wild or to use them to encourage lucrative tourism? "The objective needs to be clear," he said. "Do we want tourists or wild orangutans? The director general has to decide."

This was as close as Mr. Suprapto got to a complaint.

He went to Tanjung Puting with more funding from the government than the park had ever had before and he began to implement the policy that the ministry, on the advice of Herman Rijksen and Willie Smits and others, had determined. "When I got there," he said, "Mrs. Biruté, she treated them like people. The orangutans. She carried them, gave them 'people food' and fed them right at the camp. We took over the operations in 1990 and began to move the feeding stations farther and farther into the forest. We decided that no more orangutans should be introduced to Camp Leakey. This was a decision born out of the idea that there is potential illness in the ex-captives that could be detrimental to the whole wild population. For rehabilitation, orangutans should go to Wanariset from now on, where there is no wild population to disturb. There is room there, even now, for more orangutans."

Realizing that Mr. Suprapto was describing the exact time when Biruté and the PHPA were fighting over the Bangkok Six and the Taiwan Ten, I asked, "When did Biruté lose her permit to do research?"

"Actually the last permit that I can remember expired in 1993. But if I asked her, she'd say she has a lifelong permit. So we tolerated her presence. What could I do?"

"The OFI raises money to save orangutans. Is

it possible that it could give funds to the PHPA?"

"I don't know. It would have to go through the director general. We really wanted to work together," he said again. "But her work tends to keep the orangutans in the camp. Her staff calls them and gets them to come. Even many wild ones come into the camp. Although she's never given us any information about her staff, most of her workers have only elementary-school educations. They were out there working, but their data was not accurate or the research never really took place. There were many staff changes, too. Sometimes she would have twenty-five workers, and then some would leave because of her low wage, but we had no way of keeping up. She never gave us reports. After our policy changes, by the time I left, there were only six ex-captives hanging around the camp. The others had gone into the forest."

"Did you have trouble making your coworkers in Jakarta understand?" I asked. "Was your supervisor, for example, upset by your conflict with Biruté? Is that why you left the park?" I was imagining Mr. Sutisna in his offices at Pacific Timber.

Gracefully, Mr. Suprapto changed the direction of my question. He said, "Indonesian attitude is not to be confrontational. And to work together. So we don't say, You cannot do this. We say, Please if you think of it, don't feed the orangutans noodles and rice."

While we talked, a little girl came into the

room carrying a blanket and singing to herself. She went over to a corner and put the blanket over her head. Maybe a granddaughter, I thought to myself. "Our daughter," Mr. Suprapto said fondly. "When I went to Tanjung Puting we lost a child."

I drank my coffee, watched the little girl hiding her head, and thought about the years her father had spent in the rainforest far from his wife. I thought about the child he had left behind, who died, and I thought about the work he had tried to do in spite of the mixed messages he received from on high. This house, with its store across the front, which he stocks every weekend, is the result, I thought, as we stood up to take our leave. The day was already hot, I was already sad, and the sounds of a bird calling from a cage in the tiny courtyard did not cheer me. "Without phantasms there would be no consumers, and we'd be back with the apes," John Berger has written.

The Good Life

When I got back from my second trip to Borneo, the house being built by the Northern individualist on the alley had progressed. Its creator was hard at work in the midst of ice and sleet and snow, carving cornices. By early spring he had added two rooms to the back of the house and a door at the side. In the old part of the house, nearest the street, the window shades were still torn and the wallpaper hung in strips, but the creator had only six days in his week. During one of them, I stopped to speak to him; he was at the foot of a ladder and he put a hand up and said, "Can't you see I'm busy?"

I couldn't help thinking about the house that we had "raised" in the Dayak village. No doubt comparisons are odious, as Montaigne and my daughters have said, but I am human and comparisons are part of clambering, part of figuring things out. In the village the men had worked in concert. The elders had stayed up all night drinking and chanting to guard the house from evil spirits. It was a communal effort. I had seen the Dayak helping each other and, once the house was built, sharing the pork and rice required for a feast.

Now, in the spring chill of Toronto, I was saddened by the contrast, but I had been moody since that last glimpse of Biruté on her dock and made moodier by a trip to Miami, during which I got up my nerve and called Matthew Block, fresh out of federal prison for smuggling the Bangkok Six. I called him at World Primates, his old business, while I was attending a book fair. But I didn't really expect him to pick up the phone.

"Hello?" he'd said.

"Matthew Block?"

"Yes."

For a second I was stymied. "I'm working on a book about Borneo," I said, as if I had just come to realize the scope of my follow.

"I've never been there."

"And Biruté Galdikas."

"I've never met her."

I said that seemed very strange since Borneo had played a large part in his story and he had played a large part in Biruté's. These days whatever occurred in my life was somehow mirrored against my life there. Walking the dogs and watching a house being built. Inviting friends for dinner and remembering my Dayak hosts, who hadn't seen each other for months but who stayed up most of the night visiting with us and then rose before dawn to kill one of their four chickens for our breakfast so that on that last morning in the village, when I opened my eyes, there were feathers on the floor and small bits of chicken poaching on the fire. It's true that once I got into the speedboat, which had returned to pick us up, I felt irritable and estranged. I didn't want to stop for gas, which was inevitable, or for coffee, which was not. I didn't want any middle ground, although what remained to be done in Borneo was only a last visit to Biruté's house in Pasir Panjang, where we sat on the grass with Mr. Jacki and watched six orangutans being carried across the front yard.

I didn't want to talk about *those* six orangutans to Matthew. I wanted to talk about the six who had been found several years before in Bangkok. I said, "I wondered if you'd like to come to dinner. If you meet me here, we could walk to a place that serves good fish."

"Not in Miami. It's too dangerous."

"Eating fish?"

"Walking."

"So how did you get involved with the Bangkok Six?" I asked later, when we were sitting on a narrow rim of concrete outside the restaurant beside a line of dark water under the stars that shine on our side of the planet. It was not the sky of Borneo, and we hadn't walked, but the smell of river and the shape of things around it made me feel a little more peaceful and contented.

But Matthew didn't answer. "Everything you can imagine comes into Miami on this river," he said, as we watched a dark barge float by. And I imagined everything, or some of it, and the peaceful, contented feeling was short-lived.

I said, "You went to jail for setting the whole

thing up. But there were a couple of other people involved. Kenny Dekker in Holland. Kurt Schafer. Doesn't he run a business called Siam Farm in Thailand?" According to what I'd read about the case in the IPPL newsletters, a "network of criminals in at least seven countries organized the smuggling" of these six orangutans. Nevertheless, the IPPL has focused its energetic attacks on Block, and as he nodded, I found it hard not to feel a tinge of sympathy for him.

He told me he'd picked up strays as a child. Like Biruté. Like me. "I was always involved with animals. And birds. I care about animals," he said. "Some of the laws, not to mention the authorities who enforce them, don't do wildlife any good. A lot of animals die in the process."

I said I'd been reading about a recent sting in Chicago, during which his old friend Tony Silva was arrested for selling twelve hyacinth macaws. "The birds were in the back seat of Tony's car in brown paper bags. Like groceries or pornography."

Matthew said, "Not 'friend.' Twenty years ago he bought birds for me." Then he said that there are zoos that can't compete for the animals offered for legal sale. Zoos, who used to be in the business of selling, only exchange breeding stock now, and it's internationally controlled.

I said, "I know. I talked to one of them. But they also know every primate available . . ."

"There's always someplace that doesn't quite meet the standards," Matthew said. "If a zoo wants to buy a gorilla, and can't get one through normal channels, it resorts to the black market. What else can it do? And think about the animals! They get lost in the margins, right? They need homes. The first time I got involved with apes was seven gorillas in Cameroon. It was '84, I think. There was no place for them where they were. No cage big enough, but they couldn't go back to the trees. Like a lot of orangutans. The Cameroon deal was my first run-in with Shirley McGreal at IPPL. She doesn't like apes coming out of the wild. There was money involved — not for me — and she doesn't like that, either. But I was trying to help."

According to U.S. guidelines for federal sentencing, smuggled goods must be assigned a market value, and although orangutans sometimes sell for as much as $25,000 to $40,000 in the United States, the U.S. government agreed with Matthew's lawyers on a value of $90,000 for the six Bangkok orangutans. The lowered value meant the charges were less grave, which in turn meant Matthew was eligible for a much lighter sentence than he might have received otherwise. But Matthew provides the U.S. Army and the Centers for Disease Control with hundreds of monkeys every year for research. They need him. What he served was a sentence of thirteen months reduced to eight. Schafer, a German citizen, cooperated with the prosecutors. "I was the only one they could get," Matthew said, still insisting that he did not intend to sell illegal orangutans. "No document exists," he told me, "where anything reflects that I knew those animals were coming from anywhere but Thailand." He says he was merely the man who put buyers and sellers together, and he had nothing to do with the way the animals were shipped. "The trouble was, the government couldn't get the crates out of Bangkok, so I had no evidence to use in my own defense," he said, leaning forward a little over a plate of fish. "I'll have to live with this for the rest of my life. When I think about those animals, I feel terrible. Sometimes I can't sleep."

Kurt Schafer, according to the records, is the one who contacted Belgrade zoo director Vukosav Bojovic, who agreed to assist in the deal in exchange for two illegal siamangs. And, on February 19, 1990, when two Indonesians delivered the three crates marked live birds to the international airport in Singapore, it was Schafer who was waiting for them in the airport cafeteria, no doubt sitting with a cup of coffee and a newspaper. He's been in the business too long to look anything but casual in the midst of a deal. The world of animal trading operates by phone and fax and covert messages. The one Schafer says Matthew sent him reads like this:

302 LOCAL MESSAGE 11:28 11/24/89
ATTENTION KURT
1) PLS MAKE SURE YOU PHONE ME JUST BEFORE SENDING SHIPMENT FROM SINGA-

PORE TO MY HOUSE. SUGGEST YOU PHONE ABOUT 2–3 PM SIN TIME JUST IN CASE I RCV ANY LAST MINUTE NEWS FROM MOS. PLEASE BE CAREFUL ABT CARRYING DOCUMENTS WITH MY NAME, COPIES OF TLY ETC ON YR PERSON OR BAGS — JUST IN CASE YOU HV ANY PROBLEMS — AS WE WILL HV MORE SHIT THAN YOU — HOPEFULLY ALL GOES OK — BUT PLS BE CAREFUL ABT DOCUMENTS ALSO THE PHONES. RGDS M. PLEASE DON'T SEND TOO MANY BIRDS — FEWEST POSS.

"Your best client is the U.S. government," I said. "And who else?"

Matthew said his second-best client is Russia.

"How many monkeys do you have at the moment?" I asked. "That's what you sell, isn't it? Usually?"

"Fifteen hundred. You're welcome to visit."

The thought of those cages . . . and the future in store for the inhabitants . . . "Maybe sometime, but no thanks."

Reminding me again that he had the survival of primates foremost in his mind, he said, "Does anyone know what happened to those six orangutans? Those animals, from what I saw in the pictures, were far too small to be put back in the forest."

"What I know," Anne Russon told me later in Toronto, "is that Tanya and Bambi were both still around in '92, at the hidden feeding area at Tanjung Harapan, Biruté's secret place." Tanya, she said, was doing well, although she was sent up to Camp Leakey with a bad case of parasites. "I think it was an ex-Earthwatcher along with a drop-in tourist who figured out what was wrong and managed to medicate her," Anne said. "Tanya was always a tough cookie! She supposedly tried to drive a Mercedes through the streets of Bangkok."

Bambi, she told me, had simply disappeared. "He was listless and not wanting to eat much, though, and that's a bad sign." Then she said, "I think they were the last two alive of the six. I'd guess that Tanya had a good shot at making a success of her life; other than that, my guess would be that they're all dead."

When I asked about Fossey, the one who was reportedly "returned to the forest," Anne said, "She was also brought up from Tanjung Harapan very ill and very listless. I was told that Biruté and Charlotte took her for a walk in the woods at the end of one afternoon and put her in a tree. They claimed that she climbed up and wouldn't come down, so they eventually had to leave and come home. They went back and looked the next day and couldn't find her. But she was much too weak to have survived. Either she died and they tried to cover up the fact, or she climbed up a tree and died there."

Anne suggested I talk to Dianne Taylor-Snow. "She was the person Biruté and Shirley McGreal called in to take care of the Bangkok Six in the first place. She'd volunteered at Camp Leakey before and worked with orangutans at the San Diego zoo."

"Why didn't they call in a vet?" I asked. "Or a primatologist?"

"Because this was going to take weeks. What professional vet or primatologist can drop everything and run off to Borneo just like that?"

When I called Dianne, who had agreed to go halfway around the world to Bangkok to take charge of the suffering babies in March of 1990, and then to spend several months with them at Tanjung Puting, she did not want to talk about it. "I've never talked about what happened," she told me over the phone. "It was too personally devastating."

But she gave me some of the background. "Because of my track record, tending to sick infants at Camp Leakey round the clock, I was sent to Bangkok for one month to bring those almost-dead babies back to life. They were treating them with dog medicine."

I asked her how she had started at Camp Leakey in the first place.

Dianne said, "In 1986, I told Biruté I wanted to work with her. Because she had created a reality so believable that I desperately wanted that. She wanted cash, so I paid her, but I thought I might not like it over there. 'Give me two months,' Biruté said, 'and if you don't want to stay I'll give you your money back.' Right. I paid her to let me volunteer. I think she assessed whoever was standing in front of her and came up with a price. Different people paid different amounts. When I first went down there in '86, they were the Orangutan Research and Conservation Project, not the OFI. Later, I sat in on the first meeting of the OFI. I remember their strategy sessions — trying to project who would be in power in Indonesia so they could kiss ass. They were big on strategy. At Camp Leakey a couple of the locals called Biruté *Ibu,* and I heard Gary Shapiro say, 'We're trying to get this Ibu thing started.' I said, 'What does it mean?' He said, 'Mother Biruté.' Everyone used that title for years. In '88, when I went back, my mission was to collect blood and hair and tissue samples for Mark Stoneking at Berkeley. For that, I got phlebology training at Valley Children's Hospital and thousands of dollars' worth of donated supplies. Things like centrifuges. My husband drove me to Biruté's mother's house on Wellesley Avenue and from there I traveled with Biruté, Fred, and Jane, and five hundred kilos of luggage. What a trip! And those blood samples got Biruté in a shitload of trouble when she tried to take them out without proper permits. That's where her troubles really began."

"I keep hearing different starting points for her troubles," I said. "Tell me about the Bangkok Six. How did it end?"

I held the phone to my ear for a minute or two. Nothing. Then: "I wasn't there for the end of it.

I got sent back. I was in terrible physical shape. By the end of it, whatever they had, I had too. The babies. I had double pneumonia, anemia and starvation. We didn't have enough food."

"Why not?"

"My assistants weren't getting paid, so I was having to feed them and pay their expenses. We were downriver at Tanjung Harapan, not at Camp Leakey. Biruté didn't want us up there. But we had no way to get in supplies."

"Why'd she want you down there?"

"I think for two reasons. First, the park administrators had threatened to take control of Tanjung Harapan if Biruté didn't use it. Second, this was the first big rescue operation she'd been involved in and she had orangutan babies dropping like flies. She couldn't deal with it. She kept going by in her speedboat without stopping."

"Did she know how bad things were?"

"I kept sending up messages. 'I need help. I have sick babies. I need food!' Finally she sent Charlotte downriver with an Earthwatch team with some nurses on it. She had them draw blood from the babies. But no photographs! She didn't want anyone outside to see what was happening. Then she ordered me to destroy all my journals."

I said, "I heard from someone who worked with her that she threw all the volunteers' notes away when they left . . ."

Dianne said, "There were cartloads of notes sitting out in an old hut being eaten by roaches.

But this was different. The journals were mine."

"Did you destroy them?"

"They're in a safe place," she paused. "Along with some conversations I taped without Biruté and Charlotte knowing it. In one they both told me that if I kept my mouth shut, I'd come out of this smelling like a rose!"

"And the babies?"

"Three of them were dead when I left after two months in Borneo. Bimbo died first within one or two weeks. Ollie died in Charlotte's house while I was there. I was holding her. Those two died in my arms. Charlotte and Biruté had been force-feeding liquids into Ollie and I think they got some in her lungs. Fossey, the precious little female, was on drip-irrigation at Pasir Panjang. Then they took her up to Camp Leakey. The last day we saw Fossey she was so weak she couldn't sit up. She was three years old. She couldn't eat or drink. But Charlotte and Biruté took her for a walk into the deep forest, and when they came back a few hours later they said Fossey had escaped. They said they couldn't catch her. They left her there to die. Everybody in Camp Leakey knew it. When I confronted Charlotte, she said if I said anything about Biruté, she would beat me up. I'm not kidding. There was a big party just before I left, for the volunteers. Oh, she tries to be very Indonesian. There are always gifts and speeches at the end of Earthwatch tours. And I was there, but she never mentioned my name. Everyone there knew exactly what it meant — I did not exist. The Bangkok Six did not happen."

"Then you left?"

"I finally got to a doctor in Pangkalan Bun and he told Biruté I was so sick that I should be sent to Java immediately. He actually wrote a letter saying I was sicker than the orangutans who were dying. But I wasn't allowed to leave until Biruté could go with me. She wanted to be the first to get to a phone and get out a press release. In fact, she kept me under guard for two weeks — first at her house, then she sent me over to Charlotte's, then back up to Camp Leakey, where I had to sleep on Charlotte's floor until the Earthwatch team left. I was terribly ill. It was awful. I've never talked about these things because at least she's doing some good out there, but she kept telling everyone I was crazy. That I needed to be institutionalized, that I couldn't be believed, that I was delusional, that she had a doctor's statement about me. When we finally left, we flew to Java and checked into the Borobudur and she went straight to her room and shut the door in my face. It was then that she called Gary Shapiro and they got out the press release that she'd already prepared."

It was my turn for silence. I clung to the phone. Dianne said, "Between the two airports in Jakarta something else happened. The 'bird crates' that she'd been carrying around with her since Bangkok, that she insisted must stay with her at all times, got dumped out of the van. I couldn't figure out why she did that, and in such a weird way. This was real evidence and she just dumped it."

I said, "I heard about those crates from Matthew Block. He says he wishes he'd gone to court to fight his case but he didn't have any evidence. Why haven't you told anyone if you think she destroyed important evidence?"

"First of all, I was completely discredited for such a long time. Let's face it, she's the lady with all the power. She's got the name and the juice. If I'd screamed 'foul,' nobody would have believed me. From what I understand, Gary was the first one she called. Then the rumors started at home that I needed to be institutionalized." Dianne said it took her several years to reclaim her life, because after the rumors, no one in the world of zoos or zoology would hire her or even take her seriously.

Biruté has a home page on the Internet surrounded by pages of text provided by the OFI — pages relating to orangutan information, to Camp Leakey history, and to upcoming trips. It says, "OFI's main responsibility is to ensure that research and conservation activities at the historic Camp Leakey site continue." Six chapters of the OFI are listed along with multiple ways to give money. It is possible to become an orangutan foster parent by calling an 800 number. When I called, I was sent a list of seven orangutans. There were names and photographs.

"Where do they live?" I asked.

"In the jungle. In Borneo."

But none of them lived in the national park. "Did you ever give her my letter?" I had asked Mr. Jacki that last afternoon as we sat on the lawn in front of her house.

Mr. Jacki had said yes, but she had not handed back a reply. He had grimaced at the antics of young Fred, Biruté's son, who was roaring up and down the driveway in a Land Rover while three women strolled across the grass in front of us, each with two orangutans in her arms.

"And what do they do all day?" I had asked.

"They get carried out to the trees and they just play in them," Mr. Jacki had said.

"Do you work with Mrs. Biruté here all the time?" I had asked as the Land Rover went howling out onto the street.

"Not really. I drive her places. Or I go talk to school children about orangutans."

"What about them?"

"That they belong here. They shouldn't be killed. It's education."

"Is this funded by the OFI in the States?"

"No. We're separate."

"But they're supporting the orangutans here. They're raising a great deal of money."

"I don't know about that. Maybe to build a clinic."

"Isn't there already a new one at Tanjung Harapan?"

"The PHPA don't have any equipment. They

will be glad to have a better one in Pangkalan Bun."

"So they would bring the park orangutans here, into town?"

"Yes. I think so. I'm not really sure."

"What will happen to these backyard orangutans?" I had asked.

"They will get released in the forest."

"At Biruté's secret place?"

Mr. Jacki had looked down.

"Isn't that dangerous for the wild ones? Do you have any real quarantine here? There's no vet. You could spread TB or hepatitis . . ." I was suddenly and finally indignant.

Riska suggested that it was time for us to leave.

On the OFI Web page devoted to Somalia, the young orangutan taken from the second station, there is an update written a year after his "rescue" from the national park. It says:

> Pasir Panjang, Central Borneo, September 13, 1995.
>
> Catching a fleeting glimpse of orange, I look up through a tangle of branches. I can hardly believe my eyes; it is Somalia, all forceful, healthy thirty-eight pounds of him! We had wondered if the fragile little creature he was a year ago would survive. Rescued by Dr. Galdikas only a few days from death, Somalia was, literally, skin and bones

when I first held him. Even after a month of loving (and constant care), he weighed less than ten pounds, and clung ferociously to his foster moms.

Yet here he is, making his way through the day and the treetops as the undisputed king of this section of jungle. The infant who last year yanked my hair when I displeased him today acknowledges my presence by swinging close enough to slap my head with his hand and then, once again . . . yank my hair. Then he's off into the trees, picking berries, chasing his pal Montana and constructing nests from branches rather than the sheets and mosquito netting of last year. I suspect there will come a time soon when Somalia will no longer be coached down from the trees to be sheltered inside at night. Knowing that he is well on his way to independent survival in the forest, I feel truly fortunate to have witnessed part of his progress. Thank you. Your contributions . . . of time, and interest, and dollars in support of OFI, have saved an orangutan . . . many orangutan . . . many orangutan.

Global Positioning

I was merely trying to travel back in time, to reconcile my fantasies of Borneo with the realities of the place.

— Eric Hansen

I made a third trip to Borneo in May 1997, but this time it was different. This time, Riska and I had spent almost a year in correspondence, most of which was received by me and sent by her. Whenever a fat envelope with Indonesian stamps arrived, I knew that our weeks in the forest had not been in vain. The envelopes contained typed or handwritten pages that were growing into the moving and deeply felt story of Riska's life. With each page, she seemed to grow more sure of herself. So my third trip was made to visit her birth village and to work on *her* book as editor, but I suppose it was still part of my quest. "A mother dream," Kristin had explained, when I told her about waiting for Biruté with a baby in my arms. "Or rather, a Search for the Mother dream. That's what your quest is, isn't it?"

I am a mother and I have one, but I'm sure Kristin was right. On some level I had wanted to find *Ibu* Biruté, someone draped in suckling infants, someone I could believe in, even worship, someone to *represent* my interests.

Instead, I found myself in a rented car with a hired driver, Riska, her father, and her seven-year-old child. Everyone was carsick, and no wonder. The logging road to Kudangan village, where Riska was born, is unpaved, full of deep gullies and rocks and surrounded by the fallen carcasses of ancient trees. It looks and feels like the end of the earth, but I hid my sense of desperation and told myself I was doing something worthwhile, at least.

When we love, we love to help.

In Kudangan, I had to fight off a sudden urge to tell everyone what they should do about the coming election, the environment, their lives. Kudangan is a military outpost and it distressed me to see the army demanding payment for unwanted favors from the villagers. It distressed me to see batteries thrown in the river where we bathed, and I made a great show of taking them out and carrying them to the shore. While everyone else was washing clothes and visiting, I spent my time trying stubbornly to teach Riska's daughter, Karina, to swim or to speak to me in English. This desire to improve, improve, improve on things seemed entirely alien to my past, tolerant self, but that only increased my sense of being in the right. I didn't like the way the puppies in Riska's uncle's house were treated, or the fact that one of the neighbors had a tortoise tied to a stump. When Karina threw a tantrum one afternoon in the rented car, because she wanted the front seat, I had come to the end of my own rope and told Riska she should not let a seven-year-old child force all the adults in the car to rearrange themselves. When Riska's father moved to the crowded back without a word, I was dumbfounded. "We don't like our children to feel hurt," Riska explained.

During the long, horrible drive to Kudangan on the timber road, we had been stopped by a primitive roadblock made of oil drums and sticks of wood. It was supervised by a wild-talking man who demanded to know who we were and where we were going. It was eerie, even frightening, but what I didn't know at that point was that only a few miles west of us Dayaks and Muslim transmigrants from the island of Madura were at war. Dayaks were killing their enemies ferociously. As we bumped along over the timber road, they were brandishing severed heads and passing the ritual Red Bowl, a call to war, from village to village. Unannounced in the press, four thousand Madurese had already been killed. The barricade was simply a way to check on troubling visitors. No wonder people in the village were tense.

Kudangan, with its longhouses and rivers and trees, has a police station of sorts as well as the active army post. Of course the effect of all those guns and uniforms and outside authority is overwhelming in such a place, so, when the police and then the military came to the house where we were staying, apparently certain that I would pay them for protecting me from the dangerous Dayak who were my hosts — "Looking after me" was how they put it — I left the room and refused to answer until Riska followed and pointed out that I could make life in that village very difficult for her family. But my sense of righteousness! And outrage! Perhaps the urge for control is so inherent in my culture that the sensitivity I valued in myself had vanished.

It seemed obvious what was wrong with things. It was the damn policemen in their glossy boots. It was the enormous logging trucks with their dead loads stirring up dust on the timber road. It was the soldiers who demanded that I seek their permission for every move, even a walk to the river for a bath or a family picnic. It was the futile national election, over which the men of the village prayed. I was invited to watch as they cut the rooster's comb, sprinkled his blood on some rice, and passed around a bowl of fossils and stones, for only the gods seemed to be listening to the Dayak villagers. The people of Kudangan were whispery and frightened. "It's dangerous," Riska whispered one night from her mattress on the floor, when I asked her what was going on. "My father won't talk. Nobody wants to say anything."

After a week or so of this, it was going to be a relief to spend a few days back on the Sekonyer River in beautiful Tanjung Puting. "Headhunting preferable to this oppression by the armed of the unarmed," I wrote in my notebook. And then, indignantly, "I see Roman army marching through ancient Europe as the same process. Tree cutting, forced settlement, religious conversion, nation-state. Oh, the sight of those boys in their ugly green uniforms. Camouflage shirts! Swinging rifles as if they were toys while the men and women of the village carry their knives into the forest to bring back vegetables and fruits and possibly meat or fish."

The long trip out of Kudangan on the logging road finally silenced me. As we left, we stopped at the military post to have our papers stamped, and I sat in the car with Karina and a cousin while Riska and her father went in to pay the requisite bribes. (The way this is done is by passing an envelope full of cash under the passport or whatever while listening to a variety of insulting remarks.) When finally the urge to intervene and stop all the nonsense overcame me and I opened the car door, several rifles were instantly raised. They were aimed not at me but at the village. No wonder Biruté wants friends in high places, I thought. Then it hit me. Not respect or love. Nor even fame. It's *power* she wants. And for the next nine hours as the evidence of human "progress" unfolded, I kept my peace. We passed the great half-built steel bridge, still uncrossable, but we looked up at it from a car ferry on the water and knew it would soon make things easier for the timber companies. We followed a belching flatbed truck loaded with huge logs down the road. And there will be more, I thought. More will come.

By the time Riska and I climbed on the *Garuda* for the last time, she was no longer my guide. We were traveling now as companions. She had her book and I had mine. We were helping each other. We found Pak Herry in the PHPA office at river's edge in Kumai looking slightly worn down. Although he unrolled a map and showed me how the park had been enlarged by a third during my absence, there seemed to be a war going on over the orangutans, and he seemed to have no ammunition left. "I'm taking a leave," he said. "I may not come back." I looked at the map he touched so fondly and thought of the huge one on a wall in Toronto that I had wanted to bring as a gift. The idea seemed stupid. Pak Herry's war wasn't logistical. It was caused by an invasion of outsiders, as was the war of the Dayaks.

"How will the park get along without you?" I wondered aloud, and then to myself, If Pak Herry can't take it, who will?

Dr. Muin was still working for the PHPA, although he had been in Japan when an orangutan named Asep died and the war began. Asep was living at the second station along with little Astra, the baby who had once made Esta sick after we all held him and fed him a bottle of milk. Both of them had been "doing brilliantly," according to Emily Mason, a young British volunteer working with the PHPA. "They were always together in the forest with just occasional visits back for milk or bananas. Then Asep got sick. It may have been pneumonia. There didn't seem to be any help for him."

Riska had written to me about this: "They were always together. Then Asep died (I forgot from what kind of sickness) and about one week after that Astra died also. He got stress because he lost his companion. He did not want to eat, his weight dropped and he got diarrhea all the time."

"He was actually clinically depressed" is how Emily, a zoology student, put it. "He just lost all his will to live. It was terrible. I took care of him myself, night and day. The main thing was to get him to eat. But he got diarrhea very badly. It was constant. My skin was actually stained with it for weeks. It was at this point that Charlotte Grimm turned up to berate me for working with the PHPA. 'How could you do it?' she said. And when she saw Astra, she started screaming, 'He's going to die, he's going to die any minute! I'm a trained nurse and I'm sure of it!' Even so, I managed to keep him alive for two weeks, giving him Oralit, which was all we had, and feeding him bananas and porridge every fifteen minutes. But it wasn't enough. When he died, we buried him in the forest next to Asep."

It was when Emily was asked by the PHPA to write to a German tourist who, over several visits, had become deeply attached to Astra and

Asep that the forest where they were buried became a battleground. "This German named Ulrike Weber used to sit with Asep and Astra every day when she was here, holding them, even though this is against the park rules. At night we had to insist that she return to the Rimba. She wanted to sleep with the two babies! Sometimes she even cried when the rangers asked her to leave."

Ulrike Weber's last visit had overlapped with Emily's stay and by this time Asep and Astra were living more and more in the forest. As Asep and Astra became independent, Weber attached herself to a couple of new orangutans, again holding them and playing with them for hours at a time. This time she was staying in the cabin where Riska and I had stayed, so she could not be "sent home" at nesting time.

Then came a visit from another tourist, who told Ulrike that Biruté, the famous orangutan savior, was in Pasir Panjang. She said there was going to be a meeting at her house, and Ulrike was invited to attend. "I had to go to Sampit to get my visa renewed," Emily said, "and when I came back, the new orangutans were gone! The day Ulrike went to Biruté's house was the day they disappeared. I was convinced she had managed to pack them up and get them downriver by kelotok, so I begged the PHPA to finally go to her place, but they were kept waiting outside for two hours! In that time the babies could easily have been hidden away. Nobody wants confrontation in this country. They still have Somalia. They have lots of orangutans in that place."

Lots of orangutans.

Tempers were running high by the time I sat in Pak Herry's office and listened to his version of the story. "Oh, she is a good lobbyist," he said of Biruté. "She worked hard to get our government to give her an award recently, and now they call her a consultant for orangutans, so she wants a desk in our office. But she doesn't work for us. She wants to keep an eye on what we are doing."

As it happened, the enlarged park had swallowed all the land by Tanjung Harapan presumably held in Pak Bohap's name. In other words, it had swallowed the secret forest where Biruté had kept her orangutans after she lost her hold on Camp Leakey. Now her own backyard was all she had left.

I remembered reading about the origins of right and wrong in Frans de Waal's book *Good Natured.* He says the higher a species' level of social awareness is, the more its members realize how events ricochet through a community until they land at their own doorstep. According to de Waal, our ancestors gradually began to understand how to preserve peace and order — hence how to keep their group united against external threats — without sacrificing legitimate individual interests. They came to judge behavior that systematically undermined the social fabric as "wrong," and behavior that made a community worthwhile to live in as "right."

Didn't seem to be working on this river.

It was so different going up that last time, as if my eyes had been made, at last, to see. As if, with so little time, I could finally take in all I wanted. I noticed the strangler figs that live atop dipterocarps not as parasites but as dependent acrobats who eventually wrap so many roots around the trees they ride that they become legs, capable of supporting the fig even when the tree dies. And all that magnificence on which the orangutans depend is made possible by the tiny fig wasp. I looked up at the epiphytes. I began drawing pictures of them. Bakung. Pandanus. All of it so interlaced, so well balanced that finally I felt passionless. There was a time when *Homo sapiens* fitted in, but that time is long past.

At Tanjung Harapan, the museum was actually open. There were park T-shirts for sale, a basket or two, a few beads; Pak Herry was doing his job. The station was well repaired, clean and painted. The exhibits were nicely done, the inscriptions well written. There was a history of all the research done in the park. Biruté's name and Rod Brindamour's were there among many others who had come and gone. Suitable orangutan habitat has declined by 80 percent in the past twenty years, a sign read. Orangutan population is down by 30 to 50 percent. Then a list. On Borneo, there are somewhere between 19,000 and 30,000 orangutans. On Sumatra, between 7,000 and 11,000. Here, at Tanjung Puting, there are

1,080 to 1,800. No way, really, to count.

And already out of date.

I had asked Carey Yeager, months before, whether they had ever considered implants that would detect orangutans in the wild — the sort of thing done to polar bears. But she explained the impossibilties: first, to catch the orangutan, then to tranquilize it, then to get close enough, when it's back in the wild, to have the device transmit its whereabouts. The possibility of wounds. The expense.

"Where's Gistok?" I asked Riska now, for he was the only person of the forest I missed.

"Up at the second station. He was driving Alan crazy. But one-armed Winny, remember her? She's doing fine in the forest. She's really gone back. It's incredible! And come and see Davida. Remember how her baby died before you first came. Remember? And she was depressed? Well, she has a nice, healthy baby."

I looked up at the very tree in which I had first seen Davida sitting, disconsolate and bereft, and remembered the way I had felt back then, as if Davida could never be helped, as if what was the point of saving her to live through so much unhappiness. But she was up there now snuggling something small and hairy at her breast. I caught a glimpse and smiled. "What a nice baby," I said. And in a little while Davida came down from her tree. I was allowed to admire. And gaze. "Oh, he's beautiful," I crooned. "Good for you." The father of her baby is a wild orangutan.

I took pictures, although Davida was not in a pristine state of nature, nor could I pick her up and carry her on my hip. She was holding her own child and picking up bricks the Trekkers or someone had left near the clinic. Studying them carefully, gauging size and weight, she began building a tower. The infant suckled and she fondled it.

Upriver at Camp Leakey, Siswi and Selamat were still sitting on the dock collecting suds. It was election day and the rangers had gone downriver to vote, so there weren't any humans around to feed them. In Indonesia, everyone has to vote by law, and Riska had a difficult time getting herself a permit for the park. "I don't care about the election," she said. "What's the sense?"

With no one to feed her, I watched Siswi walk through some reeds to pick several stems of pandanus for lunch. These she ate quite happily, while Selamat stayed on the dock dangling upside down and making a fuss. He was still half-blind of course, but that was not Camp Leakey's worst tragedy: Pak Akhyar had lost the sight in both eyes. "He can't come back," I was told by Biruté's young workers the next day, who were lounging on the steps of their dilapidated quarters. "He can't see. He can't work. He went back to his village."

The young workers had caught a wild gibbon. As we talked, it huddled miserably in a cage. The three Twisted Sisters sat high in a tree. Biruté's small house looked even more like a relic. In fact, the whole of Camp Leakey looked like a piece of the past. "I never go up there," Emily had said. "I can't stand all those depressing half orangutans."

Don't take anything except foto.
Don't leave anything except trace.
Don't bring anything except memory.

We moored at the old sinking hut and saw a crocodile slide past. The moon rose and we both tried to sleep.

Then Riska began telling me a story. It was about the last part of her name. "It was the year someone first walked on the moon," she said softly. "So it means — what's that word in English . . . ?" She paused.

"Constellation." Yadi announced from the upper deck.

"Yadi speaks English?" I whispered, and thought, All this time, he's understood every word! How much had he told "Mrs. Biruté"? Why did I care? "Astronaut," I said and Riska and I laughed and tried again to sleep.

We had bathed, eaten, brushed our teeth — everything was the same. Except that it wasn't. I tried to bring back the sense of familiarity, of possessiveness, of belonging to this place. But I couldn't, I didn't. At the confluence of the small branch of the river that leads to Camp Leakey and the main artery, two waters collided — one milky with pollution, the other pure black. I had taken a picture of the two currents, pushing and tangling at the bend in different uniforms, trying to unite.

There were no other tourists on the river during that inauspicious election time, so it seemed that the forest had been given back to its "less civilized" inhabitants. As we lay on our mats in the dark, they hooted and called. But we had no shared language, no answer to send through the night. All we had was goodwill, and it was too late for that.

Domestics

Very early on the last morning, Riska and I went up to visit Trevor and Carey. Earthwatch was in residence at her station, so at last Carey's efforts to enlist them had borne fruit. But by the time we arrived, they'd all gone downriver to Kumai with Carey and Trevor on the way to West Kalimantan, where they were helping on another project. How does she do it? I wondered, already assailed by the lassitude that hits me by 10:00 A.M. in the furious heat. Slowly we climbed the hill from the dock and sat down on a bench. Except for Yadi and Carey's assistant, Lidan, no one else was around. The forest stretched out below and above. What I could see was trees, sky, Trevor's radio tower, and a sign:

> Warning: venomous snakes (cobras, kraits and pit vipers) are common.
> A common snake defence posture is to freeze.
> A snake may be mistaken for a vine or branch.

"I'm going for a walk," I said, but Riska was deep in conversation. Yadi had come up the hill with us and as they talked with Lidan, Riska started translating. "Before he began working for Carey, Lidan worked for Biruté for six years. He was there when Rico attacked Carey. He was there after that, when Rico attacked Biruté and finally got taken into the forest. First it was Carey. Then in 1983 he bit Mrs. Biruté on the right arm, and she wrapped nails in her hand so she could poke him if he came back and tried it again. Lidan is Pak Atak's brother. He says, 'With Mrs. Biruté, we had to wake up at 3:00 A.M. and go to the forest on a follow for twelve to fifteen hours five days a week.' He says, 'Here we work only in the morning when it is cool. In the afternoon we can rest.' "

Good, I thought, good. But all I want is to climb the tower. Then, out of my unconscious or the thicket above the shore, I heard something calling. "Hornbill!" Riska said. Once I had heard a hornbill passing with such a beating of wings, such inflating and deflating of air pouches, that it sounded like a steam engine going past. But never before had I heard the great bird call out.

Inside the legs of the tower, wooden steps snaked back and forth to the top, which was not very high, but from which I could see trees that stretched all the way to heaven, the way an ocean does when you are in the midst of it. The female hornbill builds her nest high in a tree, sealing herself in so there's no escape. While she sits on her eggs, her mate brings all her food, poking it in through the tiny hole she has left in the mud and twigs. Such trust, I thought, while in the air nothing moved except a thousand invisible things. Life. Heat. The voices of three *Homo sapiens*. "A lot of people are coming here, more than a hundred," Lidan was saying. "They're making a movie of Mrs. Biruté's life."

Tanjung Puting — the national park — had increased in size since our second visit, but would it ever be large enough?

"Who will be Mrs. Biruté?" Yadi asked.

"A movie star. Isabella Rosie . . ."

"Rossellini," Riska guessed.

Where will they sleep? I wondered. How will they eat? A movie shoot means a lot of people. Couldn't they do it somewhere else? Biruté came to Tanjung Puting all those years ago to study orangutans. But right away she ran into a problem. She was trying to study a species that was being threatened by human beings. The rainforest was being cut down to make room for palm plantations. For agriculture. For mining and settlements. It was being cut down because logging was an obvious way to get rich. And everyone was doing it, from the president of the country right down to the local politicians. The main job of the Ministry of Forestry was to hand out timber concessions, and that was a nice job to have. Foreign companies wanted the concessions and paid huge bribes to get them. The foreign businesses had, by law, to be connected to Indonesians, and that made a lot more people rich.

But an orangutan needs a lot of forest.

According to the signs in the new museum at

Tanjung Harapan, because fruit trees are widely dispersed, one orangutan needs five hundred hectares (1,200 acres) just to survive. According to the signs, this place differs from other lowland, mixed dipterocarp forests in Borneo because a great part of it stands in black water, steeping in acids, alkaloids, and pigments that have leached out from the toxic leaf litter. But whereas there may be 35 native species of tree in all of England, there are 780 right here. And after logging they never come back, not really. After logging there is the sun and the rain unimpeded. Erosion. After logging only imperata grass will grow, and it is too coarse even for grazing. What then? Tourists and filmmakers are better by half, obviously.

In Toronto, a friend of mine has a new kind of map on her wall. Beth runs a mining exploration company. "And you should see what a group of us bought," she had told me a few weeks before, over lunch. "It's a satellite view of Kalimantan. I mean every grain of sand. It's a miracle what we can see from space. Technology has revolutionized the mining industry."

This is the map I had wanted for Pak Herry's office wall. "Could I get a copy of the map for the national park?"

Beth had looked embarrassed.

"I mean, how much would something like this cost?" I had pictured myself unrolling the large, multicolored map in that faraway riverside office.

"Two hundred thousand dollars." Beth's embarrassment was on my behalf. "But you can come see it."

The map was beautiful. Silvery skin against black night, its undulations — remarkable only on the closest inspection — represent the alluvial curves of topsoil and sand under rainforest and rivers. Beth's miracle of technology can decode the elements in each speck. "With that and a GPS we can put a helicopter down within five meters of any particular place." I think of mats made of grass that has been buried and dried. Prayers offered on bamboo wings. The human flight of a hornbill dance. The value given to relics. But Indonesia is made of magmatic arcs. There are trenches in the earth where molten rock is pouring out, forming land, and many of them are full of minerals: copper, silver, gold. Kalimantan means "river of precious stones." And so it is. A map can tell where the gold is likely to be, where the hardwood trees are likely to be, and even where orangutans are likely to be, but it can't tell a thing about how to help them coexist with us.

If I had been able to see around the curve of the earth from Trevor's tower, I might have looked into the future and seen the fires that were bound to start, the fires that were bound to burn and burn in Tanjung Puting and all over Borneo for months. Location, location, location.

"The good news about working with Indonesia," Beth reports, "is that anything is made possible by money. The bad news is that anything is made possible by money. The government is entirely corrupt, and they don't give a damn about damage to the environment. Anything we do right in that direction is up to us. There is always a palm to be crossed, but once that's done, things can really get done. Things can get done and stay done. And if you can get your payment all the way up to the first family, everything will get done right. A business can't always come up with the requirements of other places, but it can always come up with cash."

By summer, those businesses would clear away the forest by burning it down on a scale no Dayak could imagine. By fall the smoke from those burning trees would pollute the atmosphere over most of south Asia and certain other parts of the world. By fall orangutans would begin to come down from the dangerous trees and flee to human gardens and fields. Where they would be attacked. Where the infants who survived the murder of their mothers would be sold to poachers, or turned over to the first available "authorities" — sometimes soldiers, sometimes police. Sometimes Ibu Biruté at her home in Pasir Panjang. By fall newspapers everywhere would be full of concerned editorials. But the bad news about working in Indonesia is that Suharto and his friends and relatives turned the forest into a cash machine. By winter his government and its economy would have collapsed, but for the species exiled from the forest for good, no election, no intervention by the IMF, was going to make things right. Rainforests and the rivers that run through them have supported human and nonhuman populations for millions of years without any serious imbalance. Until now.

Biruté had told the students in Long Beach that Jakarta is very much like Los Angeles, "both with smog and traffic," she said. "And this is where decisions concerning the tropical rainforests of Indonesia are made." Of course, they are made in Washington, too. They are made in Toronto. And Tokyo. But a few of those decisions were still being made in the forest, where the "orangutan war" was raging. *And this, too,* I thought, as Riska and I traveled back down the Sekonyer, *has been one of the dark places of the earth.*

When we left Carey's station, Riska and I seemed to have nowhere left to go except down to the second station on our way out of the park. I wanted to see Gistok. I was sure he'd be waiting. But it was Emily Mason, the young volunteer, who was sitting on the dock, and we invited her to come aboard and sit in the shade and have lunch with us. She'd been living in Tanjung Puting for eight months and knew all the workers as well as all the orangutans. "What's the population these days?" I asked, as soon as she had settled on a pillow and tucked in her bare feet.

"Tanjung Harapan, ten," she rattled off.

"Here at Pondok Tanggui, twelve. Camp Leakey, five, really, with another forty who return occasionally."

"Successful rehabs?"

"Pandok Tanggui, four. Two at Camp Leakey."

Emily has long blond hair and an English face. She was wearing a T-shirt and sarong and said she never wanted to leave Tanjung Puting. During her sojourn, she'd become fluent in the local dialect as well as Bahasa Indonesia.

"And Gistok? Where is he? I thought he'd come down to meet the boat out of curiosity."

"Not anymore. Since he came up here he sticks to the trees. Oh, maybe he comes down to grab a banana or two, but for the most part he lives in the forest full-time."

"You're kidding."

"No, I'm not. He even sleeps there. He can build a nest."

I took a walk up the dock. Talked to the rangers. Looked around. No Gistok. Making a little bow when we left, I tipped my cap to him in absentia. He had moved away like any other child. A domestic species can survive anyplace, and perhaps Gistok is one of us now. We *Homo sapiens*, according to John Livingston of York University, first domesticated *ourselves*. Then we domesticated dogs, cows, horses . . . maybe even orangutans. And a domestic species can survive anyplace, but it has no place in particular to call its own. (Inside the cap was this tag: "Embroidered with pride in Englewood, Colorado. Made in China.") Gistok is one of us now, except that we are more adaptable. And there is a problem with our adaptability, because wherever we go we make things difficult for local species. When species evolve side by side, snakes and frogs, predators and prey in the same place, they achieve a nice equilibrium. But no predators adapted along with us, growing the right kind of fangs or beaks to keep our numbers down. Just look at us!

Even my peony waits three years to bloom each time I transplant it. An odd balance, this — possibly the best insurance biodiversity could have except that leaving us out of it means it was achieved at a terrible cost. We have overrun the planet and killed off hundreds of thousands of species. Species who cohabit successfully with us, be they dogs, cats, or seagulls, tend to overpopulate. The city pigeon, the rat . . . and where does that leave a fellow like Gistok? If he can live off our garbage, if he can live in the shadows of our lumber camps and gold mines, he'll survive. He likes rice. He likes clothes. He likes lipstick. Eventually his descendants, if he manages to have any and if they learn to get along with us, may even wipe out their wild cousins in the forest the way we have wiped out ours. They'll throw a few diseases into the trees or multiply to such an extent that there isn't enough food or room for all the others.

We were once natural beings. But no longer.

We have built a great wall of culture, of ideas and information, between us and our home. Without the wall, we would starve or freeze or die of loneliness. Without the wall, we would perish. In fact, a few weeks later I heard that Gistok was back, breaking into the rangers' kitchen and being "domestic" to such an extent that he was finally locked up in a cage. I fear it may have an orangutan-proof latch.

After we left the second station we stopped one more time before we got to Kumai. Across the river from Tanjung Harapan, in front of the village where Pak Atak lives, someone has built a small lodge. Pak Herry had suggested we take a look at it, but I had my doubts. The owner may have worked with the PHPA and the villagers, but he is from Java and I suspected that he'd imported the staff from home and would send all the profits there. What we found, when we climbed up on the narrow, well-built dock, was an eight-room hotel built in the local style and almost invisible in the landscape of trees. The rooms are better than attractive; they're simple, handcrafted, and quite beautiful. The dining room is small. The staff are each and every one hired from the village of Tanjung Harapan, and the mats they weave provide color on the clean wood floors. I was glad to see Pak Atak's carvings on sale in the dining room and to visit him in his new house with his wife and child. The village looked ever so slightly more prosperous. I bought a basket and three small mats. I greeted the people I recognized.

But soon, nipa and pandanus closed around us again and we began the short trip out of the rainforest for the last time together. We were two women this time, and we had found a real balance, both of us having made Rousseau's journeys, one to a place where life is uncorrupted (for I believe that can still be said of the first Dayak village) and another into the self.

For many months after that last trip up the Sekonyer, I tried to discover, by sifting the evidence, what had changed Biruté's path. Why had her forest become a jungle? Was it character or circumstance?

"My heart goes out to her," Earthwatch Europe's director Andrew Mitchell told a reporter shortly after the organization's break with Biruté. "It is the same pattern as Dian Fossey. She became a scientist, but she has become more and more attached to the animals and more involved in conflicts with local people and authorities. I fear that she is shipwrecking herself."

Fossey died a martyr, had a movie made of her life, had numerous and vociferous detractors (including Biruté), and was compared, by the man who delivered her eulogy, to Jesus Christ. She was assumed to be mad, paranoid, alcoholic, delusional, enormously daring, and vastly insecure.

Jane Goodall is the good fairy of Leakey's

three angels, universally admired and adored. But the three women, like sisters intertwined, have borne constant comparison to each other. Tied to the other two by Louis Leakey, Biruté has had to live in the great shadows they cast.

Through nearly a decade of turmoil, Biruté projected an air of serene confidence, but that time has passed. In 1998 ecotourists hugged the half-tame and often sickly orangutans by the feeding platforms of Tanjung Puting while loggers ignored the endangered status of the trees and the creatures that live in them, and a pall of smoke hung over the Indonesian archipelago for a second year. Finally, as the survival of our forest cousins became more conjectural with each passing day, the government of Indonesia called for an investigation into Biruté's activities in Pasir Panjang.

In October 1997 the Ministry of Forestry had received a letter from one of Biruté's volunteers, a schoolteacher from California. Michelle Desilets first went to Borneo on an OFI tour in 1994 and then went back to work with the orangutans at Birute's house six times over the next three years. The letter she sent to the ministry concerned her experiences during the summer of 1997, shortly after my return to Canada, when the forest had begun to burn and the orangutans had begun to come down from the high canopy.

The first part echoes, very closely, the story I had heard from Dianne Taylor-Snow. It's a story of house arrest, of stolen journals and photographs, of personal threats. What must have been more surprising to the ministry, however, were Michelle's allegations about the numbers of orangutans kept illegally at Biruté's house, and her certainty that Biruté had paid for some of them and stolen others from the national park. She told of a home "nursery" in Pasir Panjang filled beyond capacity, of orangutans left alone in cages, of medical neglect, of bedding "saturated with diarrhea and worms," of too little food, too little medicine, too little milk.

"There is much concern about a number of orangutans either disappearing or dying and no explanation is given," she wrote. "There is much concern about large amounts of money coming in from team members and paying volunteers that does not appear to be supporting the orangutans. There is concern about Galdikas charging Dutch students $8,000 to do research in the government-run Tanjung Puting National Park and charging some volunteers $1,000 per month when absolutely nothing is provided."

According to Michelle, there had been sixty-five orangutans on Biruté's premises by the end of the summer, including "large, healthy orangutans such as Somalia, Henny and Massru still being housed at Galdikas' home. Why have they not been returned to the wild?"

In the middle of a month of bloody food riots and an investigation of corruption charges leveled at Suharto, the Ministry of Forestry declared that, "having been in the forefront of orangutan conservation, Dr. Biruté Galdikas should be expected to be aware of Indonesian rules and regulations and to continuously set an example to abide by the laws."

When I read Michelle Desilets's report several months later, three lines, embedded somewhere in the middle, stood out irrevocably. They say simply that "Galdikas solicited the advice of a zoo veterinarian, Dr. Sasha, on the care of Siswi. The Dr. stipulated that Siswi should be left alone when not receiving injections to lower anxiety. This advice was ignored." For me, learning that Siswi had been stolen from Camp Leakey was the last, terrible straw. If mine is a story of searching for a mother, it's about searching for daughters as well — mine and Biruté's. About needing our daughters so fiercely that we put that need first, before even their welfare. (When Siswi tried to go back to the forest, she was locked with her half-blind child in Birutés attic!)

When I read that report, I remembered that, in Kumai as we were negotiating for a taxi to take us to Pangkalan Bun, Riska ran into a friend of hers, another Dayak who had trained with her to be a guide. "He's an old friend of Pak Bohap's. He helps them confiscate orangutans."

"Mrs. Biruté, she's *bagus* [best]," the old friend said, thumbs up, as if he knew he was living on a battlefield. As if he'd chosen his side.

"And you go out with her to confiscate orangutans?"

"We got two last time around two months ago." He shrugged, then smiled.

We said good-bye to Yadi, to his parents, and to his wife and child. I took photographs. A car had been found. Climbing into it, I looked back at the waterside. Sadly. Because, if anything divides us from our fellow creatures, it is our fragile ability to know ourselves as outside of paradise.

Space, why be afraid of it? What if we lose every tree? There will still be a planet with or without us and with or without orangutans. Where before I looked at a dark blue spruce every day from my kitchen, I now look into my neighbors' house, where there are several orchids in terracotta pots on glass shelves behind glass. We are closer now than we were before. I watch them eat dinner, two men in shirtsleeves. I watch one of them leave each morning wearing a tie. I listen to their laughter on summer evenings and try to think of us as villagers. Houses, first dwellings, caves; first sense of awe recorded there; first place where we became domestic — and thereby turned ourselves into what we are. *Homo sapiens sapiens*.

I think of the moment Gistok walked up and put his hand around mine. It felt like the baseball glove worn by an older brother every afternoon through childhood, warm, already shaped, already knowing how to grip, pulling me to the trees, leading me away from those who were staring after us. And I felt the reluctance I feel with anyone who is very determined, as if I'd prefer to choose my own time. But I went, and Gistok walked me around the cabin I would live in a few months later and we peered underneath, where there was nothing to see but the bare ground on which he had slept. And all of what happened was ahead of me, mine to live in for a time.

EPILOGUE

In February 1998, an inspection team was formed by the Inspector General in Jakarta and the Director General of the PHPA, and on March 3 they went to the Galdikas house in Pasir Panjang. A report of that visit includes the following observations:

> Based on the results of this assessment, at this time in the house of Dr. Biruté Galdikas were found 89 (eighty-nine) orangutans, which were held in four secret lodges placed in the forest behind the house. By coincidence, the joint team also met with a foreigner (who identified herself as a tourist from Denmark) named Ms. Sigy who was carrying 3 (three) young orangutans. She admitted that she had already lived for two weeks in the house of Dr. Biruté Galdikas, to help take care of sick orangutans. Because it was getting late into the evening, the team asked to meet with Ms. Sigy the next morning for an interview, but when that morning came she had left.

The investigators' later interviews were more fruitful:

> Based on the confessions of Ms. Walyati, private secretary, and Dr. Ichlas Al Zaqie, staff researcher, and corroborated by Mr. Bohap Bin Jalan, husband of Dr. Biruté Galdikas, the orangutans present were obtained from Tanjung Puting National Park, Palangka Raya, and Pangkalan Bun, and the rest were handed over by local people.

Galdikas, the report went on to say, paid local people between 80,000 and 300,000 rupiah ($10–28) for each orangutan.

The investigators then described what they saw when they went into the "secret lodges":

> The maintenance of orangutans takes place in a system of cages, in which the orangutans described are placed in a pen measuring 1 meter × 1.5 meters, occupied by 2–5 orangutans, then each morning are fed, removed from the pen, and brought by a guardian to be released. In the afternoon the orangutans are taken and given food and brought back and placed into their pens.
>
> For the care of sick orangutans there is no quarantine or place for medical treatment. In the case of an orangutan that becomes sick, for treatment it is taken to a doctor for humans (not a veterinarian), because the majority of diseases that orangutans suffer are also suffered by humans. In addition to this, from time to time, visiting foreign doc-

tors provided free medical treatment.

The money for the care of the orangutans came from Dr. Biruté Galdikas privately, as well as from the Orangutan Foundation International and from the province of Central Kalimantan, in the sum of 500,000 rupiah [$46] per month. In addition to this funding, according to admission of a former volunteer worker who worked with Dr. Biruté Galdikas, each foreign volunteer worker and researcher is required to pay $1,000 [U.S.] per month.

The condition of the place for accommodating the orangutans does not meet health standards. Isolation cages were made of metal measuring 1.5 meters × 1 meter × 1 meter. One such pen would be occupied by 3–5 orangutans. The isolation cages are placed in wooden sheds measuring 10 meters × 4 meters, each holding approximately 10 isolation cages. At the time of the inspection, the floors of the sheds [were] covered with fruit peels and feces, including diarrhea. Two of the sheds were located close to chicken pens owned by other people, and there were two dogs roaming the sheds. . . . Wandering around were several baby orangutans with diarrhea. In one pen occupied by 3–5 baby orangutans, these babies were not free to move about, and their cages also had feces in them. Three young orangutans were found in a hut without ventilation and light, being cared for by a German tourist. One baby orangutan was feverish, while another was wearing a diaper for its diarrhea.

Based on the results of the Inspection Team's observations, even adult orangutans revealed dependence on humans; this was observed at the moment of release from the pens when they hang on to and beg to be carried by the caregiver. Furthermore, those that are still babies like even more to be carried by the caregiver. The presence of tourists to care for sick baby orangutans is illegal because the care of sick baby orangutans by non-medical personnel can endanger the health of these orangutans.

Based on the testimony of Dr. Biruté Galdikas's former volunteer workers, at least twenty-one orangutans died because of insufficient care or misdiagnosis. Although the volunteer workers said the orangutans were buried behind the kitchen, the Inspection Team could not prove this claim because it is outside the competence of the Team [the uncovering of graves and autopsies].

Included in the report is a review of Galdikas's plan to build an orangutan clinic. In February 1996, the OFI and the Indonesian government had agreed to cooperate in the building of an Orangutan Care Center that would become the property of the Indonesian government but would be run jointly with the OFI, which would provide $500,000 (U.S.). The memorandum of understanding, signed by the government and the OFI, included the government in every step of the planning and building of the project. It stated that "the Orangutan Foundation International and its personnel under this Memorandum of Understanding will not engage in political affairs and any ventures or activities in Indonesia." Now, two years later, the inspection team reported back on the center's progress:

> Dr. Biruté Galdikas took unilateral actions (not in accordance with the Operation Plan which was agreed to by SETKAB), this fact is evidenced by:
>
> The decision of the location for the development of the care center was determined by Dr. Biruté Galdikas alone, on land she owned that is located in Pasir Panjang village, Pangkalan Bun, without weighing the proposal and advice of the head of Tanjung Puting National Park.
>
> The appointment of the contractor CV Bumi Lambung Mangkurat Pangkalan Bun was made by Dr. Biruté Galdikas as though she were the head of the Management and Quarantine of Orangutans project, without the knowledge of the head of Tanjung Puting National Park. Based on the field inspection, the development of the care center is

approximately thirty percent complete.

The inspection committee made a number of recommendations:

> The Director General PHPA needs to immediately take control of the activities of Dr. Galdikas, both her activities in Tanjung Puting National Park and those of rearing and caring for orangutans in her private home.

In a letter dated April 17, 1998, the Director General of the PHPA informed Galdikas that by holding orangutans at her house she was in violation of the law; that they were being held in substandard and illegal conditions; that she had violated their memorandum of understanding by beginning work on the Orangutan Care Center; and that she should "stop the activities of constructing the Primate Care Center."

On May 15, 1998, Galdikas responded:

> Concerning the inspection results by the Team that at least eighty-nine orangutans were around the house of my husband, an indigenous from Central Kalimantan. At the time of the inspection by the Inspection Team there were fifty-six orangutans, which we had reported to the Minister of Forestry and the PHPA Director General. These orangutans we had received from people/

police officials and from the Head of Tanjung Puting National Park himself. Because these orangutans were in a state of illness and lack of care, and suffered additional casualties from the forest fires of August through November 1997, we happily accepted the delivery of these orangutans for care.

We cared for these orangutans and they were able to live in the forest and former primary jungle which is as wide as +/–0.36 hectares [0.9 acres] as a habitat located near our Temporary Care and Quarantine, for the orangutans which we manage. These orangutans live safely, because humans cannot enter into this habitat, such that the orangutans are not contaminated with human diseases. Villagers in the area participate in protecting it, as there are wild fruit trees in this forest. In this habitat, the orangutans receive exercise where wild orangutan had previously lived, but have now run away. . . . Our raising of orangutans is appropriate within our function as Consultant, which was established by the Minister of Forestry. Besides this, we wish to express our complaint and disappointment in the attitude of the Inspection Team, which used rude conduct in entering people's homes.

If there are issues that are unclear, as Consultant appointed by the Minister, it is only fitting that we come to give full and detailed explanations. We should report that we employ 70 employees, the majority of whom are villagers who live in the area of the Temporary Quarantine and Care Center, and among whom we employ 4 biology graduates.

Regarding the construction of the Orangutan Care Center Clinic we report the following:

We had difficulty meeting with the Head of Tanjung Puting National Park for consultation on the completion of the Construction of the Center for Care and Quarantine of Orangutans, so we relied on the oral agreement of the Director General PHPA, made during five meetings. And after this, we intended to report to the Head of Tanjung Puting National Park, but until now we have not yet been able to meet. We have tried to speak on the telephone with him and have only succeeded in connecting with him two times, during which in these calls the Head of Tanjung Puting National Park gave his agreement for the Construction of the Orangutan Care and Quarantine Center described above.

Regarding the funding which has emerged as a result of the Cooperative Team between OFI and the Director General PHPA based on the MOU, estimations which have been charged to OFI and which we should report are that the funds from OFI have been sent directly to the Contractor's Bank Account, have never been sent through us, and we have never used foreign financial assistance or any foundation for our personal self-interest. At the same time, the funds for the administrative and technical salaries have never been realized, because of this we request instruction on the method for realizing release of the above funds.

Regarding the construction of the Orangutan Care and Quarantine Center, we will be faithful to Director General PHPA's instructions in accordance with his letter dated 19 August 1997, concerning continued cooperative OFI-PHPA action.

A month later, Indonesian officials monitoring the situation discovered that the construction on the care center had not stopped, and that, in fact, the center was almost complete. Shortly afterward, promotional literature for the OFI announced that, "With approval of the Indonesian government, OFI built an orangutan care center and quarantine for confiscated orangutans needing to be medically treated and cared for in preparation for reintroduction to the wild." These developments, along with continued concern about Galdikas's orangutans, angered some PHPA officials. An internal PHPA memo dated June 4, 1998, reported that the total number of orangutans at Galdikas's house remained unchanged at eighty-nine, and that several of the orangutans ran away and were found on the main road.

> We do not agree or we are not of the opinion whatsoever that the rehabilitation activities of orangutans [should] continue to take place in Dr. Biruté Galdikas's home, in Pasir Panjang; even if all the funding for it comes from Dr. Biruté Galdikas/OFI. The reason is that we fear that this will give the impression that the PHPA legalizes these illegal activities conducted by Dr. Biruté Galdikas.

The memo recommended placing strict controls on Galdikas's ability to conduct research and rehabilitation. It also recommended that no formal title be given to Galdikas again, and concluded by suggesting that the PHPA request a formal, written apology from Galdikas addressed to the Indonesian government.

The Director General of the PHPA then drafted a letter to Galdikas, sent on June 15, 1998, which said in part:

> You are charged with reporting in writing and with specific details the total number of orangutans that have been cared for and died in your home.

Adding that:

> The construction of the Primate Care Cen-

ter which is being constructed in Pasir Panjang Village is to be stopped.

And finally:

The statement in point 2 of your letter is not correct. The Inspection Team at the time of entering your home behaved politely and were well received by Mr. Bohap and were welcomed to inspect the pens of orangutans.

To date, it is unclear whether the condition of relations between Galdikas and the Indonesian government has improved, or whether the orangutans have been removed from Galdikas's house. The Orangutan Foundation International still runs its research tours to Tanjung Puting.

Acknowledgments

When Susan Renouf at Key Porter Books first approached me with the idea of this book, neither of us imagined the friendships (especially our own) that would develop from the experience. So first, I thank Susan for her inspiration and unflagging support and for providing the many friends I was soon to meet. One of them is Dr. Anne Russon, from the Department of Psychology at York University, who shared her extensive library, her own research into the behavior of orangutans, and her considerable knowledge of the Tanjung Puting area. I want to thank Anne for help without which the book could not have been written.

I thank my mother, Edith Senner Dickinson, to whom the book is dedicated, for handing on her curiosity about the natural world, her delight in it, her own culture. I thank, most particularly, my daughters, Esta and Kristin, for agreeing to share the adventure and for participating so fully in the responsibilities our trip entailed, which included close reading of the manuscript afterward, and without whose companionship and emotional support the follow would not have happened. I thank Riska Orpa Sari, who led me into the rainforest and safely out again on three occasions, and whose extensive knowledge of that forest and its diverse inhabitants brought meaning and focus to an experience that was otherwise too intense, too unfamiliar, to record fairly. Not only did Riska join the physical follow, she joined the spirit of it as well, becoming friend and confidante, sending me briefings and updates whenever she visited Tanjung Puting, and connecting me to her own very private world. I thank, always, Michael Ondaatje, who held things together at home in my absence, who encouraged me throughout, who read the manuscript and advised me in its structuring, and who, unfailingly, brought light to an often dark subject. And I thank Douglas Fudge, who read the manuscript as a scientist and provided a very helpful edit.

I appreciate the help of Dr. Herry Sjoko Susilo, head of Tanjung Puting National Park, who made it possible for me to live there for an extended time, who spent several of his extremely busy work hours explaining its routines and problems, and who has enhanced the accessibility of the park through informative exhibits and his own pleasant personality. Dr. Carey Yeager and Trevor Blondal were immensely helpful in providing advice, arranging for boat and shelter, allowing us to visit their research station and inviting Esta to join a tree-measuring expedition. Further, their interviews were extensive and quite important to the book, and I am grateful for their continued work in the forest and for permission to quote our conversations. I thank Dr. Abdul Muin for the materials he provided and for his candor and help, and Mr. Suprapto, former head of the park, who kindly invited me into his home in Jakarta and spoke to me at length. Yadi, I thank for his unflagging good humor, his skill with a kelotok, and his patience with me. Kate Linbaugh, for her assistance in Jakarta; Elizabeth Kirkwood, Heather Brazier, and Constance Russell for their help and research. Gloria Bishop for her reading of the manuscript. Laurie Coulter and Duncan Murrell for sensitive editing. To Lyn Miles, Emily Mason, and Willie Smits, thanks go for sharing both facts and insights.

In Los Angeles, David and Liz Ondaatje and Jacob Dickinson provided guidance, shelter, and cheer. In Bali, two people were particularly helpful: Suzanne Baron, who showed us the island while describing its culture unforgettably, and Hans Iluk, who has led many trips of his own into Tanjung Puting and is a strong advocate for Borneo's surviving orangutans. It was Hans who put me in touch with Riska and with Shirley McGreal, head of the International Primate Protection League, who has been of invaluable help in decoding the legal tangles of the Bangkok Six as well as providing background on the Taiwan Ten.

I appreciate the help of Kathy Glassco in making contact with Matthew Block and joining us during our interview in Miami. I thank Matthew

Block and Dianne Taylor-Snow for sharing the history of the Bangkok Six. I thank Michelle Desilets, and Michelle Molina, founders of Primate Survival Coalition, each of whom made a great leap of faith in agreeing to talk to me. Pak Atak in Tanjung Harapan and Hajji Makmur, wherever he is, I thank for speaking so candidly. Pak Akhyar I thank for his kindness and for an introduction to the specifics of Dayak knowledge. Pak Pangi, I thank for the same, as well as Yefni's parents, who graciously opened their home to me for a five-day stay.

I want, finally, to thank Gistok, Siswi, and Davida, who taught me a great many things, but especially that all of this business of saving is nothing if not an exchange.

The employees of G.K. Hall hope you have enjoyed this Large Print book. All our Large Print titles are designed for easy reading, and all our books are made to last. Other G.K. Hall books are available at your library, through selected bookstores, or directly from us.

For information about titles, please call:

(800) 257-5157

To share your comments, please write:

Publisher
G.K. Hall & Co.
P.O. Box 159
Thorndike, ME 04986